Finite Element Methods

MECHANICAL ENGINEERING

A Series of Textbooks and Reference Books

EDITORS

L. L. FAULKNER
Department of Mechanical Engineering
The Ohio State University
Columbus, Ohio

S. B. MENKES
Department of Mechanical Engineering
The City College of the
City University of New York
New York, New York

OTHER VOLUMES IN PREPARATION

Finite Element Methods

An Introduction

RONALD L. HUSTON
University of Cincinnati
Cincinnati, Ohio

CHRIS E. PASSERELLO
Michigan Technological University
Houghton, Michigan

MARCEL DEKKER, INC.   New York and Basel

Library of Congress Cataloging in Publication Data

Huston, Ronald L.
  Finite element methods.

  (Mechanical engineering; 25)
  Includes index.
  1. Finite element method. I. Passerello, Chris E.,
[date]. II. Title. III. Series.
TA347.F5H87   1984   620'.001'515353   83-20876
ISBN 0-8247-7070-6

MARCEL DEKKER, INC.
270 Madison Avenue, New York, New York 10016

Current printing (last digit):
10 9 8 7 6 5 4 3 2

PRINTED IN THE UNITED STATES OF AMERICA

Preface

This text is the outgrowth of lecture notes given in a three-quarter introductory finite element course at the University of Cincinnati for a number of years. It is intended primarily for advanced undergraduate and beginning graduate students in engineering and the physical sciences. It is also hoped that the practicing engineer will be able to learn the subject through independent study of the text.

The objective of the text is to introduce the reader to the terminology and procedures of the finite element method so that he or she may be able to confidently pursue advanced studies in the subject and/or use the method in the solution of particular problems. More specifically, upon completion of the text, the reader should be able to write a finite element computer program, confidently use the standard industrial finite element programs, and read the ever-increasing research papers and books on the subject.

The organization of the text generally follows the historical development of the subject. It begins with a brief review of the necessary mathematical techniques, which may be omitted or used as a reference as needed. This is followed by an introduction to the subject through one-dimensional elements applied to trusses and frames. The subject matter is then further developed through two-dimensional elements applied to heat transfer, plane stress, and plane strain. The text concludes with a brief introduction to iso-parametric elements. Problems are provided at the end of each chapter. Finally, the text contains example problems and algorithms applicable in each of the above-mentioned areas.

The authors are grateful to many students and others for their help in preparation of the manuscript. Special thanks are given to Ms. Janet Gaffney, Mrs. Grace Martin, and Mrs. Jeannette Ghandi for typing the manuscript and for help in correspondence. Thanks are also given to Mr. Dimitrios Mavriplis and Mr. Majid Hasan for help in preparing the figures. Finally, the authors appreciate the advice and helpful suggestions of Dr. James Kamman of Notre Dame University, who carefully reviewed the entire manuscript.

Ronald L. Huston
Chris E. Passerello

Contents

Finite Element Methods

1

Introduction

1.1 THE FINITE ELEMENT METHOD

Finite element methods and techniques have become established, and in many instances essential, procedures for engineers and scientists. In this text our objective is to introduce the fundamental operations and procedures associated with finite element methods. Our approach will be to introduce and develop the procedures through simple yet increasingly complex examples.

The finite element method basically consists of the following procedures: First, a given physical or mathematical problem is modelled by dividing it into small or fundamental parts called "elements." Next, an analysis of the physics or mathematics of the problem is made on these elements. Finally, the elements are reassembled into the whole with the solution to the original problem obtained through this assembly procedure.

Although finite element methods are currently being applied in many fields of science and engineering, our emphasis in this text will be with problems in structural mechanics. Indeed, we will study trusses, beams and frames, and plane elastic problems. However, to show that the basic procedures of the method are applicable in other areas, we will also apply the method with problems in heat transfer and with differential equations.

Before we begin our discussion of the finite elements procedures however, it will probably be helpful to review a few ideas from applied mathematics which are used in the development of these procedures. We will conduct this review in the following sections of this introductory chapter. We will assume that the reader is already familiar with these ideas and that he or she has access to texts providing more rigor and detail than would be appropriate here.

Specifically, we will briefly discuss matrix methods, variational techniques, Lagrange interpolation, and Hermitian interpolation. A short bibliography is provided at the end of the chapter to help the reader find additional information about these ideas and about the finite element method itself.

1.2   MATRIX METHODS

Recall that a matrix A is a rectangular array of numbers a_{ij} ($i = 1$, $\ldots$, m; $j = 1$, $\ldots$, n) with m rows and n columns written in the form

$$A = [a_{ij}] = \begin{bmatrix} a_{11} & a_{12} & \cdots & a_{1n} \\ a_{21} & a_{22} & \cdots & a_{2n} \\ \cdots & \cdots & \cdots & \cdots \\ a_{m1} & a_{m2} & \cdots & a_{mn} \end{bmatrix} \tag{1.1}$$

The numbers a_{ij} are usually called the "elements" of the matrix, but to avoid confusion with the terminology of finite element methods, we will refer to them in this text as "entries" or "components." The subscripts i and j of a_{ij} refer to the row and column in which a_{ij} occurs.

Two matrices are said to be *equal* if they have identical or equal entries. That is,

$$A = B, \quad \text{when} \quad a_{ij} = b_{ij} \ (i = 1, \ldots, m; j = 1, \ldots, n) \tag{1.2}$$

If all the entries of a matrix A are zero, the matrix is called a *zero matrix* and hence,

$$A = 0, \quad \text{when} \quad a_{ij} = 0 \ (i = 1, \ldots, m; j = 1, \ldots, n) \quad (1.3)$$

If a matrix has only one row, it is called a *row matrix*. If a matrix has only one column, it is called a *column matrix*. If a matrix has an equal number of rows and columns, it is called a *square matrix*. If a square matrix A has all zero entries except for the diagonal entries, it is called a *diagonal matrix*. If all the diagonal entries of a diagonal matrix are ones, the matrix is called the *identity matrix* and is generally give the symbol I. That is,

$$A = I, \quad \text{when} \quad a_{ij} = \delta_{ij} = \begin{cases} 1 & i = j \\ 0 & i \neq j \end{cases} \quad (1.4)$$

where δ_{ij} is defined by the expression and is often called "Kronecker's delta symbol." The *transpose* of a matrix A denoted by A^T is defined as the matrix obtained by interchanging the rows and columns of A. Hence, for a matrix A with m rows and n columns and entries a_{ij}, the transpose A^T has n rows and m columns and entries a_{ji}. For a square matrix, if $A = A^T$, the matrix is said to be *symmetric*. If $A = -A^T$, the matrix is said to be *antisymmetric*. Finally, for a square matrix, it is possible to evaluate the *determinant* of the matrix. (The reader should review the elementary properties and operations of determinants if needed.) If the determinant of a matrix is zero, the matrix is said to be a *singular* matrix.

The algebra of matrices is based upon a few relatively simple rules: That is, the multiplication of a matrix A by a scalar s produces a matrix B whose entries b_{ij} are simply sa_{ij}, where a_{ij} are the entries of A. That is,

$$sA = B, \quad \text{where} \quad b_{ij} = sa_{ij} \ (i = 1, \ldots, m; j = 1, \ldots, n) \quad (1.5)$$

If two matrices A and B each have m rows and n columns, they may be added. The sum, written as $A + B$, is the matrix C whose entries are the respective sums of the entries of A and B. That is,

$$A + B = C, \quad \text{where} \quad c_{ij} = a_{ij} + b_{ij}$$
$$(i = 1, \ldots, m; j = 1, \ldots, n) \quad (1.6)$$

Subtraction is defined similarly. If two matrices A and B are such that the number of columns n of A is equal to the number of rows m of B, the matrices A and B are said to be *conformable*. Conformable matrices may be multiplied. The product of two conformable matrices A and B is defined through the "row-column-product" procedure. That is, if C is the product of A and B, then a typical entry c_{ij} of C is the sum of the products of the respective entries of the ith row of A with the jth row of B. Specifically, if

$$C = AB, \quad \text{then} \quad c_{ij} = \sum_{k=1}^{n} a_{ik} b_{kj} \tag{1.7}$$

It is easily shown that multiplication is associative but not commutative. That is,

$$ABC = (AB)C = A(BC) \tag{1.8}$$

but, in general

$$AB \neq BA \tag{1.9}$$

Also, multiplication is distributive over addition and subtraction. That is,

$$A(B + C) = AB + AC \tag{1.10}$$

Finally, it is easily shown that the transpose of a product of two matrices A and B is the product of their transposes in reverse order. That is,

$$(AB)^T = B^T A^T \tag{1.11}$$

If A is a nonsingular square matrix, it is possible to define the *inverse* of A, written as A^{-1}, as the matrix with the property that the product of A and A^{-1} in either order is the identity matrix. That is,

$$AA^{-1} = A^{-1}A = I \tag{1.12}$$

A^{-1} may be determined as follows: Let $\tilde{A}_{ij}$ be the "adjoint" of the entry a_{ij}. (Recall that the adjoint A_{ij} of the entry a_{ij} is defined as

$$\tilde{A}_{ij} = (-1)^{i+j} M_{ij} \qquad (1.13)$$

where M_{ij} is the "minor" of a_{ij} defined as the determinant of the matrix A with the ith row and the jth column deleted.) Then, A^{-1} is given by

$$A^{-1} = [\tilde{A}_{ij}]^T / \det A \qquad (1.14)$$

where $\det A$ represents the determinant of A.

If a square matrix A has the property that its inverse A^{-1} is equal to its transpose A^T, then A is said to be *orthogonal*. (We will have occasion to use orthogonal matrices in a number of finite element applications.)

A useful procedure in matrix multiplication is *partitioning*. Partitioning is simply the division of a matrix A into *submatrices*. For example, the 4 X 5 (4 rows and 5 columns) matrix A might be partitioned as follows:

$$A = \begin{bmatrix} a_{11} & a_{12} & | & a_{13} & a_{14} & | & a_{15} \\ a_{21} & a_{22} & | & a_{23} & a_{24} & | & a_{25} \\ \hline a_{31} & a_{32} & | & a_{33} & a_{34} & | & a_{35} \\ a_{41} & a_{42} & | & a_{43} & a_{44} & | & a_{45} \end{bmatrix} = \begin{bmatrix} A_{11} & | & A_{12} & | & A_{13} \\ \hline A_{21} & | & A_{22} & | & A_{23} \end{bmatrix} \qquad (1.15)$$

where the submatrices A_{ij} ($i = 1, 2; j = 1, 2, 3$) are defined by inspection. If two matrices A and B are partitioned such that their submatrices are appropriately conformable, the submatrices may be treated as entries and then the product of A and B may be represented in terms of the products of the submatrices. For example, for the 5 X 3 matrix B partitioned as

$$B = \begin{bmatrix} b_{11} & b_{12} & b_{13} \\ b_{21} & b_{22} & b_{23} \\ \hline b_{31} & b_{32} & b_{33} \\ \hline b_{41} & b_{42} & b_{43} \\ \hline b_{51} & b_{52} & b_{53} \end{bmatrix} = \begin{bmatrix} B_{11} \\ \hline B_{21} \\ \hline B_{31} \end{bmatrix} \qquad (1.16)$$

and for the matrix A of Equation (1.15), the product AB may be expressed as:

$$AB = \left[\begin{array}{c} A_{11}B_{11} + A_{12}B_{21} + A_{13}B_{31} \\ \hline A_{21}B_{11} + A_{22}B_{21} + A_{23}B_{31} \end{array} \right] \qquad (1.17)$$

We conclude this brief review of matrix methods by mentioning a second (but not as well known) matrix procedure which we will find especially helpful with our finite element analysis. It is called *expansion* and it consists of simply increasing the number of rows and/or columns in a matrix, or a matrix equation, by inserting rows and/or columns of zeros. For example, consider the following matrix equation:

$$\begin{bmatrix} a & b & c & d \\ e & f & g & h \\ i & j & k & l \end{bmatrix} = \begin{bmatrix} 8 & 2 & 9 & -3 \\ 4 & -1 & 6 & 4 \\ 5 & -6 & -7 & 1 \end{bmatrix} \qquad (1.18)$$

This equation is, of course, equivalent to the 12 scalar equations [see Equation (1.2)]:

$$a = 8, \quad b = 2, \quad c = 9, \quad d = -3, \quad e = 4, \quad f = -1$$
$$\qquad (1.19)$$
$$g = 6, \quad h = 4, \quad i = 5, \quad j = -6 \quad k = -7, \ l = 1$$

By expanding the size of the matrices of Equation (1.18) by inserting rows and columns of zeros respectively, leads to a matrix equation which is still equivalent to the same set of scalar equations as in Equations (1.19). For example, the matrices may be expanded as follows:

$$\begin{bmatrix} a & 0 & 0 & b & 0 & c & 0 & d \\ 0 & 0 & 0 & 0 & 0 & 0 & 0 & 0 \\ 0 & 0 & 0 & 0 & 0 & 0 & 0 & 0 \\ e & 0 & 0 & f & 0 & g & 0 & h \\ i & 0 & 0 & j & 0 & k & 0 & 1 \end{bmatrix} = \begin{bmatrix} 8 & 0 & 0 & 2 & 0 & 9 & 0 & -3 \\ 0 & 0 & 0 & 0 & 0 & 0 & 0 & 0 \\ 0 & 0 & 0 & 0 & 0 & 0 & 0 & 0 \\ 4 & 0 & 0 & -1 & 0 & 6 & 0 & 4 \\ 5 & 0 & 0 & -6 & 0 & -7 & 0 & 1 \end{bmatrix}$$
$$\qquad (1.20)$$

1.3   VARIATIONAL TECHNIQUES

Variational techniques play a central role in the formulation of the fundamental governing equations in finite element methods. Therefore, we will present a brief review of these techniques here. As before, the reader is, of course, encouraged to consult other references for more rigor and details.

Recall that the basic problem stimulating the development of variational calculus is to find a function $y(x)$ such that the integral I given by

$$I = \int_a^b F(x, y, y')\,dx \tag{1.21}$$

has a maximum or minimum value where $y(a) = c$, $y(b) = d$, and y' is dy/dx. Recall also that the desired function $y(x)$ is obtained by introducing a "variation" $\epsilon\eta(x)$ and adding it to the $y(x)$ appearing in Equation (1.21). $\eta(x)$ is an integrable but otherwise arbitrary function and ϵ is a small parameter.

By replacing y by $y + \epsilon\eta$ in Equation (1.21), I may be written in the form

$$I = I_m + \delta I = \int_a^b F(x, y + \epsilon\eta, y' + \epsilon\eta')\,dx \tag{1.22}$$

where I_m is the desired maximum or minimum value of I and δI is the "variation of I" due to the variation $\epsilon\eta$ of y. In Equation (1.22), $y(x)$ has become the desired maximizing or minimizing function. By expanding $F(x, y + \epsilon\eta, y' + \epsilon\eta')$ in a Taylor series in ϵ and by retaining only the linear terms, we obtain

$$F(x, y + \epsilon\eta, y' + \epsilon\eta') = F(x, y, y') + \epsilon[(\partial F/\partial y)\eta + (\partial F/\partial y')\eta'] \tag{1.23}$$

By substituting this expression into Equation (1.22), δI becomes

$$\delta I = \epsilon \int_a^b [(\partial F/\partial y)\eta + (\partial F/\partial y')\eta']\,dx \tag{1.24}$$

By integrating the second term by parts, we obtain

$$\delta I = \epsilon\eta(\partial F/\partial y')\Big|_a^b + \epsilon \int_a^b \eta[(\partial F/\partial y) - d(\partial F/\partial y')/dx]\, dx \tag{1.25}$$

Now, if either η or $\partial F/\partial y'$ is zero at a and b, δI becomes

$$\delta I = \epsilon \int_a^b \eta[(\partial F/\partial y) - d(\partial F/\partial y')/dx]\, dx \tag{1.26}$$

Finally, by observing in Equation (1.22) that I is I_m when δI is zero and by noting that $\eta(x)$ is arbitrary, we see that the integrand is zero for all $\eta(x)$ if

$$d(\partial F/\partial y')/dx - \partial F/\partial y = 0 \tag{1.27}$$

Equation (1.27) is commonly called the "Euler-Lagrange equation." If F does not depend explicitly on x, Equation (1.27) may be integrated leading to the expression

$$F - y'\,(\partial F/\partial y') = \text{constant} \tag{1.28}$$

A convenience often employed in these kinds of analyses is the introduction of the "variational operator." That is, the "variation" δ is often introduced and defined as

$$\delta F = \left(\frac{\partial F}{\partial \epsilon}\bigg|_{\epsilon=0}\right)\epsilon \tag{1.29}$$

It is easily shown that the operator δ possesses most of the properties of differential operators (for example, the sum rule, the product rule, etc.). Using this notation, it is readily seen from Equation (1.29) that

$$\delta y = \epsilon\eta \tag{1.30}$$

By beginning with Equation (1.21) and following similar steps to those in Equations (1.24) to (1.26), we find that (see Problem 1.6)

$$\delta I = \int_a^b [(\partial F/\partial y) - d(\partial F/\partial y')/dx]\, \delta y\, dx = 0 \tag{1.31}$$

For a very simple illustration of these ideas, consider the following problem: What is the shape of the curve connecting the two points $P_1(x_1, y_1)$ and $P_2(x_2, y_2)$ which possesses the shortest length? To solve this problem, we simply form an expression for the length L in the form of the integral

$$L = \int_{x_1}^{x_2} [1 + (dy/dx)^2]^{1/2} \, dx \qquad (1.32)$$

The problem is then simply to find $y(x)$ such that L is a minimum. By identifying the integrand in Equation (1.32) with $F(x, y, y')$ of Equation (1.28), we obtain

$$[1 + (y')^2]^{1/2} - (y')^2/[1 + (y')^2]^{1/2} = c \text{ (a constant)}$$

or

$$1/[1 + (y')^2]^{1/2} = c \qquad (1.33)$$

Solving for y', we obtain

$$y' = c/\sqrt{1 - c^2} = m \text{ (a constant)} \qquad (1.34)$$

Then

$$y = mx + b \qquad (1.35)$$

where b is a constant. This is, of course, the familiar equation of a straight line. By adjusting m and b so that the line passes through P_1 and P_2, we obtain, finally,

$$y = y_1 + [(y_2 - y_1)/(x_2 - x_1)](x - x_1) \qquad (1.36)$$

For a second illustration, consider the integral

$$I = \int_a^b [(y')^2 + y^2 + 2yf] \, dx \qquad (1.37)$$

where $f = f(x)$ is a given function. Applying the Euler-Lagrange Equation (1.27) to the integrand immediately leads to the equation

$$d^2y/dx^2 - y = f(x) \qquad (1.38)$$

In this case we see that the problem of solving the differential equation in Equation (1.38) is equivalent to minimizing the integral in Equation (1.37). Indeed, the variational procedure as described above and as developed in the Euler-Lagrange Equation (1.27) can be viewed as a procedure for converting the minimization problem into a differential equation. However, sometimes it may be desirable to reverse this procedure. That is, there may be occasions when we will want to convert a differential equation into a minimization problem. Such occasions often arise in finite element analyses.

The procedure for converting a differential equation into an integral extremum problem is, in general, relatively simple and it can be considered as the reverse of the procedure used in the development of the Euler-Lagrange equation. Specifically, we can reverse the procedure by simply multiplying the differential equation by a variation of the dependent variable δy and then integrating by parts. To briefly illustrate this, consider again the differential equation of Equation (1.38), rewritten

$$d^2y/dx^2 - y - f = 0 \qquad (1.39)$$

If we multiply by δy and integrate over the interval (a, b), we obtain

$$\int_a^b [d^2y/dx^2 - y - f] \delta y \, dx = 0 \qquad (1.40)$$

If we integrate the first term by parts, we obtain

$$\int_a^b (d^2y/dx^2) \delta y \, dx = (dy/dx) \delta y \Big|_a^b - \int_a^b y' \delta y' \, dx \qquad (1.41)$$

where in view of Equation (1.29) we have interchanged the operations of differentiation and variation and where, as before, y' is dy/dx. If δy or dy/dx is zero at a and b, Equation (1.41) may be rewritten

$$\int_a^b (d^2y/dx^2) \delta y \, dx = -\int_a^b y' \delta y' \, dx = -\tfrac{1}{2} \int_a^b \delta(y')^2 \, dx \qquad (1.42)$$

The second and third terms of the integral of Equation (1.40) may be written

$$\int_a^b y\delta y\, dx = \frac{1}{2}\int_a^b \delta y^2\, dx \tag{1.43}$$

and

$$\int_a^b f\delta y\, dx = \frac{1}{2}\int_a^b \delta(2fy)\, dx \tag{1.44}$$

(Note that $\delta f = 0$ since f does not depend explicitly upon y.) By combining the results of Equations (1.42), (1.43), and (1.44), we see that the integral of Equation (1.40) may be written

$$\int_a^b [d^2y/dx^2 - y - f]\,\delta y\, dx = -\frac{1}{2}\int_a^b \delta[(y')^2 + y^2 + 2yf]\, dx \tag{1.45}$$

or simply as

$$\delta I = 0 \tag{1.46}$$

where I is given by Equation (1.37). Equation (1.46) represents the desired integral minimization problem.

As a final example, and one which will also be useful in the sequel, consider applying the above procedure with the differential equation of beam theory

$$EI\, d^4v/dx^4 = p(x) \tag{1.47}$$

where x is along the beam axis, v represents the transverse deflection, $p(x)$ is a given loading function, and EI is a flexural constant. By multiplying by δv and by integrating between 0 and ℓ, we obtain

$$\int_0^\ell [EI\,(d^4v/dx^4) - p(x)]\,\delta v\, dx = 0 \tag{1.48}$$

where ℓ is the beam length. By integrating the first term by parts, we obtain

$$\int_0^\ell EI(d^4v/dx^4)\delta v\ dx\ =\ EI\frac{d^3v}{dx^3}\ \delta y\Big|_0^\ell - \int_0^\ell EI(d^3v/dx^3)\delta v'\ dx$$

$$(1.49)$$

By integrating by parts again, we obtain

$$\int_0^\ell EI(d^4v/dx^4)\delta v\ dx\ =\ EI\frac{d^3v}{dx^3}\ \delta v\Big|_0^\ell - EI\frac{d^2y}{dx^2}\ \delta v'\Big|_0^\ell$$

$$+ \int_0^\ell EI\ v''\ \delta v''\ dx \qquad (1.50)$$

If v and dv/dx are unspecified (for example, unknown) at 0 and ℓ, it is appropriate to let δv and $\delta v'$ be unspecified. Also, if we adopt the notation that

$$f_1\ =\ EI\ d^3v/dx^3\Big|_{x=0}\ =\ EI\ v'''\ (0) \qquad (1.51)$$

$$f_2\ =\ -EI\ d^2y/dx^2\Big|_{x=0}\ =\ -EI\ v''\ (0) \qquad (1.52)$$

$$f_3\ =\ -EI\ d^3v/dx^3\Big|_{x=\ell}\ =\ -EI\ v'''\ (\ell) \qquad (1.53)$$

and

$$f_4\ =\ EI\ d^2v/dx^2\Big|_{x=\ell}\ =\ EI\ v''\ (\ell) \qquad (1.54)$$

then, the first term in the integral of Equation (1.48) may be written

$$\int_0^\ell EI(d^4v/dx^4)\delta v\ dx\ =\ \tfrac{1}{2}\int_0^\ell EI\ \delta(v'')^2\ dx - f_1\ \delta v(0) - f_2\ \delta v'(0)$$

$$- f_3\ \delta v(\ell) - f_4\ \delta v'(\ell) \qquad (1.55)$$

Hence, if we define I as

$$I = \tfrac{1}{2} \int_0^{\ell} [EI(d^2v/dx^2)^2 - 2vf]\, dx - f_1 v\,(0) - f_2 v'\,(0)$$
$$- f_3 v\,(\ell) - f_4 v'\,(\ell) \tag{1.56}$$

then the beam differential equation of Equation (1.47) is seen to be equivalent to minimizing I. That is

$$\delta I = 0 \tag{1.57}$$

1.4 LAGRANGE INTERPOLATION

Recall that the development of the Lagrange interpolation polynomials grows out of the solution of the following problem: Given a set of $n + 1$ points in the X-Y plane, find the polynomial of degree n which passes through these points. Specifically, consider the $n + 1$ points shown in Figure 1.1. Let $P(x)$ be the desired polynomial of degree n which passes through these points. The problem, of course, is to find $P(x)$. The solution is usually written in the form

$$P(x) = y_0 L_0(x) + y_1 L_1(x) + \cdots + y_n L_n(x) \tag{1.58}$$

where the functions $L_0(x)$, $L_1(x)$, ..., $L_n(x)$ are polynomials of degree n. The task is then to find the polynomials $L_i(x)$ ($i = 0, \ldots, n$). These polynomials are called the "Lagrange polynomials."

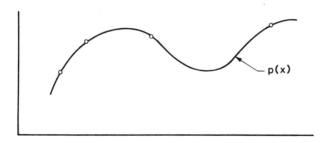

Figure 1.1   A Polynomial $P(x)$ Passing Through
a Given Set of n Points.

At this point, it might be helpful to consider the following remarks: First, the polynomial $P(x)$ is unique (see Problem 1.8.). That means that if we can find *a* polynomial which passes through the points, it is *the* desired polynomial. Second, since a sum of polynomials of degree n is also a polynomial of degree n, $P(x)$ of Equation (1.58) is a polynomial of degree n. Finally, since $P(x)$ is unique and since $P(x)$ must pass through each of the n points, it is seen from Equation (1.58) that, in general, the $L_i(x)$ must have the properties

$$L_i(x_j) = \begin{cases} 0 & i \neq j \\ 1 & i = j \end{cases} \quad (i, j = 0, \ldots, n) \tag{1.59}$$

That is,

$$P(x_j) = y_j \quad (j = 0, \ldots, n) \tag{1.60}$$

The properties defined in Equation (1.60) enable us to determine the $L_i(x)$: Since $L_i(x_j) = 0$ when $i \neq j$, the x_j $(j = 0, \ldots, n)$, $(j \neq i)$ are *roots* of $L_i(x)$. Hence, $L_i(x)$ may be written in the form

$$L_i(x) = C_i(x - x_0)(x - x_1) \cdots (x - x_{i-1})(x - x_{i+1}) \cdots (x - x_n) \tag{1.61}$$

where C_i is a constant. C_i may be determined by observing that since $L_i(x_i) = 1$, we have

$$1 = C_i(x_i - x_0)(x_i - x_1) \cdots (x_i - x_{i-1})(x_i - x_{i+1}) \cdots (x_i - x_n) \tag{1.62}$$

Hence, $L_i(x)$ takes the form

$$L_i(x) = \frac{(x - x_0)(x - x_1) \cdots (x - x_{i-1})(x - x_{i+1}) \cdots (x - x_n)}{(x_i - x_0)(x_i - x_1) \cdots (x_i - x_{i-1})(x_i - x_{i+1}) \cdots (x_i - x_n)} \tag{1.63}$$

Using the product notation Π, this may be written in the more compact form

$$L_i(x) = \prod_{j=0}^{n} \frac{(x - x_j)}{(x_i - x_j)} \tag{1.64}$$

Finally, $P(x)$ takes the form

$$P(x) = \frac{(x - x_1)(x - x_2) \cdots (x - x_n)}{(x_0 - x_1)(x_0 - x_2) \cdots (x_0 - x_n)} y_0$$

$$+ \frac{(x - x_0)(x - x_2) \cdots (x - x_n)}{(x_1 - x_0)(x_1 - x_2) \cdots (x_1 - x_n)} y_1$$

$$+ \cdots + \frac{(x - x_0)(x - x_1) \cdots (x - x_{n-1})}{(x_n - x_0)(x_n - x_1) \cdots (x_n - x_{n-1})} y_n \tag{1.65}$$

Using both summation and product notations, Σ and Π, this in turn may be written in the compact form

$$P(x) = \sum_{i=0}^{n} \prod_{\substack{j=0 \\ j \neq i}}^{n} \frac{(x - x_j)}{(x_i - x_j)} y_i \tag{1.66}$$

Equation (1.66) is useful in many areas of numerical computation including both interpolation and integration. (See Problems 1.9, 1.10, and 1.11.) However, the simplicity of the Lagrange polynomials, especially when they are written in the compact form of Equation (1.64), can be deceptive. Indeed, if $n \geqslant 3$ the analysis can be quite cumbersome. Fortunately for our purposes in finite element techniques it will be sufficient to consider only the linear and quadratic terms. That is, if $P(x)$ is quadratic (a parabola passing through three points), it takes the form

$$P(x) = \frac{(x - x_1)(x - x_2)}{(x_0 - x_1)(x_0 - x_2)} y_0 + \frac{(x - x_0)(x - x_2)}{(x_1 - x_0)(x_1 - x_2)} y_1$$

$$+ \frac{(x - x_0)(x - x_1)}{(x_2 - x_0)(x_2 - x_1)} y_2 \tag{1.67}$$

Finally, if $P(x)$ is linear (a straight line passing through two points), it takes the form

$$P(x) = \frac{(x - x_1)}{(x_0 - x_1)} y_0 + \frac{(x - x_0)}{(x_1 - x_0)} y_1 \qquad (1.68)$$

Equations (1.67) and (1.68) will be useful in the sequel.

1.5   HERMITIAN INTERPOLATION

The development of the expressions for the Hermitian interpolating polynomials follows the same pattern as that for the Lagrange polynomials. That is, a polynomial is sought which passes through a given set of points, but in addition, the polynomial must also satisfy derivative values at these points.

Since this is an extension of the requirements for the Lagrange polynomials and since the Lagrange polynomials can involve tedious and even cumbersome expressions, the analysis is even more involved with Hermite polynomials. Therefore, we will restrict our discussion to only the very simple cases. Specifically, we will consider seeking a polynomial which passes through two points with the first derivatives specified at these points. That is, let y_1 and y_1' be specified at x_1 and let y_2 and y_2' be specified at x_2 as indicated in Figure 1.2. Since four conditions are specified (that is, y_1, y_1', y_2, and y_2'), the polynomial which will uniquely satisfy these conditions will be third order. As with the Lagrange polynomials, we will express the desired polynomial in the form [see Equation (1.58)]

$$P(x) = y_1 h_1(x) + y_1' h_2(x) + y_2 h_3(x) + y_2' h_4(x) \qquad (1.69)$$

where $h_1(x)$, $h_2(x)$, $h_3(x)$, and $h_4(x)$ are each third order, or cubic, polynomials. Moreover, they have the properties

$$h_1(x_1) = 1, \quad h_1'(x_1) = h_1(x_2) = h_1'(x_2) = 0 \qquad (1.70)$$

$$h_2'(x_1) = 1, \quad h_2(x_1) = h_2(x_2) = h_2'(x_2) = 0 \qquad (1.71)$$

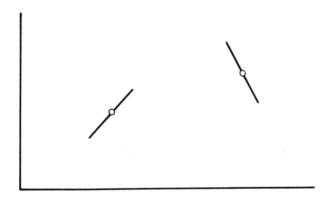

Figure 1.2   Two Given Points and Slopes

$$h_3(x_2) = 1, \quad h_3(x_1) = h'_3(x_1) = h'_3(x_2) = 0 \qquad (1.72)$$

and

$$h'_4(x_2) = 1, \quad h_4(x_1) = h'_4(x'_1) = h_4(x_2) = 0 \qquad (1.73)$$

It is relatively easy to show that the polynomials satisfying these conditions are (see Problem 1.11)

$$h_1(x) = (x - x_2)^2 \left[(x_1 - x_2) + 2(x_1 - x)\right]/(x_1 - x_2)^3 \qquad (1.74)$$

$$h_2(x) = (x - x_1)(x - x_2)^2/(x_1 - x_2)^2 \qquad (1.75)$$

$$h_3(x) = (x - x_1)^2 \left[(x_2 - x_1) + 2(x_2 - x)\right]/(x_2 - x_1)^3 \qquad (1.76)$$

and

$$h_4(x) = (x - x_1)^2 (x - x_2)/(x_1 - x_2)^2 \qquad (1.77)$$

These polynomials are called "cubic Hermitian interpolates" and they will be very useful to us in the sequel.

PROBLEMS

Section 1.2

1.1   Review the rules for evaluating determinants.

1.2   Verify Equations (1.8), (1.9), (1.10), and (1.11).

1.3   Verify Equation (1.12) using the definition of A^{-1} in Equation (1.14).

1.4   Show that the matrix A given by

$$A = \begin{bmatrix} \cos\theta & \sin\theta & 0 \\ -\sin\theta & \cos\theta & 0 \\ 0 & 0 & 1 \end{bmatrix}$$

is orthogonal.

1.5   Verify Equation (1.17) by direct multiplication of the matrices of Equations (1.15) and (1.16).

Section 1.3

1.6   Verify Equation (1.31) by taking the variation of I in Equation (1.21) and by integrating by parts.

1.7   Find the curve passing through the points $P_1(x_1, y_1)$ and $P_2(x_2, y_2)$ in the X-Y plane which generates the smallest area for a surface of revolution when rotated about the X-axis.

Answer: $y = c_1 \cosh[(x/c_1) + c_2]$

where c_1 and c_2 are determined by requiring the curve to pass through P_1 and P_2.

Hint: Recall that the area of a surface of revolution may be expressed as:

$$A = 2\pi \int_{x_1}^{x_2} y[1 + (dy/dx)^2]^{\frac{1}{2}}\, dx$$

Section 1.4

1.8 Show that a polynomial $p(x)$ of degree n passing through $n + 1$ points is uniquely determined.

Hint: Let $p(x)$ have the form

$$p(x) = a_0 + a_1 x + a_2 x^2 + \cdots + a_n x^n$$
$$= b_0 + b_1 x + b_2 x^2 + \cdots + b_n x^n$$

Let $c_i = a_i - b_i$ $(i = 0, \ldots, n)$. Then, by letting $x = x_i$ $(i = 0, \ldots, n)$, develop $n + 1$ linear equations for the $n + 1$ c_i.

1.9 Given the following table of values or coordinates:

X	0	1	2	3
Y	3	4	7	6

Find an expression for the polynomial which passes through the points having these coordinates.

Solution:

$$P(x) = 3L_0(x) + 4L_1(x) + 7L_2(x) + 6L_3(x)$$

where

$$L_0(x) = (-x^3 + 6x^2 - 11x + 6)/6$$
$$L_1(x) = (x^3 - 5x^2 + 6x)/2$$
$$L_2(x) = (-x^3 + 4x^2 - 3x)/2$$
$$L_3(x) = (x^3 - 3x^2 + 2x)/6$$

1.10 Using the results of Problem 1.9, interpolate the data of the table in that problem to find the value of y which corresponds to $x = 1.3$.

1.11 (a) Observe from Equation (1.66) that the area under the polynomial of Figure 1.1 between x_0 and x_n may be expressed in the form

$$A = \int_{x_0}^{x_n} \left[\sum_{i=0}^{n} \prod_{\substack{j=0 \\ i \neq j}}^{n} \frac{(x - x_j)}{(x_i - x_j)} \, y_i \right] dx$$

(b) Observe that a more efficient approach to obtaining the area might be obtained by dividing the region into N *elements* each of which contains three points as shown in Figure P1.11b. That is, the area may be expressed as

$$A = \sum_{e=1}^{N} \overset{(e)}{A}$$

where $\overset{(e)}{A}$ is the area of a typical element.

(c) Consider a typical element (e). Let the coordinates of the three points of this element be $(\overset{(e)}{x_1}, \overset{(e)}{y_1})$, $(\overset{(e)}{x_2}, \overset{(e)}{y_2})$, and $(\overset{(e)}{x_3}, \overset{(e)}{y_3})$ as shown in

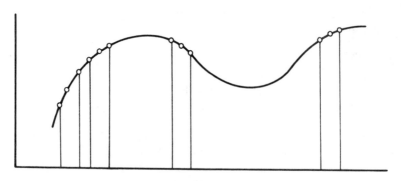

Figure P1.11b   Region of Figure 1.1 Divided into N Elements.

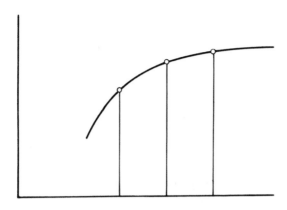

Figure P1.11c   Typical Element (e).

Figure P1.11c. Show that the polynomial passing through these three points may be expressed as

$$\overset{(e)}{p}(x) = \overset{(e)}{y_1}\overset{(e)}{L_1}(x) + \overset{(e)}{y_2}\overset{(e)}{L_2}(x) + \overset{(e)}{y_3}\overset{(e)}{L_3}(x)$$

where $\overset{(e)}{L_1}(x)$, $\overset{(e)}{L_2}(x)$, and $\overset{(e)}{L_3}(x)$ are

$$\overset{(e)}{L_1}(x) = \frac{(x - \overset{(e)}{x_2})(x - \overset{(e)}{x_3})}{(\overset{(e)}{x_1} - \overset{(e)}{x_2})(\overset{(e)}{x_1} - \overset{(e)}{x_3})}$$

$$\overset{(e)}{L_2}(x) = \frac{(x - \overset{(e)}{x_1})(x - \overset{(e)}{x_3})}{(\overset{(e)}{x_2} - \overset{(e)}{x_1})(\overset{(e)}{x_2} - \overset{(e)}{x_3})}$$

$$\overset{(e)}{L_3}(x) = \frac{(x - \overset{(e)}{x_1})(x - \overset{(e)}{x_2})}{(\overset{(e)}{x_3} - \overset{(e)}{x_1})(\overset{(e)}{x_3} - \overset{(e)}{x_2})}$$

(d) Let the horizontal separation between the points be h as shown in Figure P1.11c. Show that the area of a typical element $A^{(e)}$ can then be expressed as

$$A^{(e)} = \int_{x_1^{(e)}}^{x_1^{(e)}+2h} p(x)\,dx = \int_{x_1^{(e)}}^{x_1^{(e)}+2h} \left[y_1^{(e)} L_1^{(e)}(x) + y_2^{(e)} L_2^{(e)}(x) + y_3^{(e)} L_3^{(e)}(x) \right] dx$$

(e) Finally, evaluate $A^{(e)}$ and thus derive Simpson's one-third rule.

$$\text{Answer:}\quad A^{(e)} = (h/3)\,(y_1^{(e)} + 4y_2^{(e)} + y_3^{(e)})$$

Suggestion: Observe that there is no loss in generality by letting $x_1^{(e)} = -h$, $x_2^{(e)} = 0$ and $x_3^{(e)} = h$. Hence, integrate the expression in part d) between $-h$ and h.

Section 1.5

1.11 Verify Equations (1.74) to (1.77).

Suggestion: Let the desired interpolate $h(x)$ be expressed in the form

$$h(x) = c_0 + c_1(x - x_1) + c_2(x - x_1)^2 + c_3(x - x_1)^3$$

and let $\ell = x_2 - x_1$. Then, by successively satisfying the conditions of Equations (1.70) to (1.73), c_1, c_2, c_3, and c_4 are determined.

BIBLIOGRAPHY

I. Applied Mathematics: Matrix Methods, Computational Methods and Variational Techniques.
    Acton, F. S., *Numerical Methods that Work*, Harper and Row, 1970.
    Ayres, F., Jr., *Theory and Problems of Matrices*, Schaum-McGraw Hill, 1962.

Forray, M. J., *Variational Calculus in Science and Engineering*, McGraw Hill, 1968.

Hildebrand, F. B., *Advanced Calculus for Applications*, Prentice Hall, 1964.

Hildebrand, F. B., *Methods of Applied Mathematics*, Prentice Hall, 1965.

Hohn, F. E., *Elementary Matrix Algebra*, Macmillan, 1966.

Ketler, R. L., and Prowel, S. P., Jr., *Modern Methods of Engineering Computation*, McGraw Hill, 1969.

Kreyzig, E., *Advanced Engineering Mathematics*, Wiley, 1963.

Kunz, K. S., *Numerical Analysis*, McGraw Hill, 1957.

Prenter, P. M., *Splines and Variational Methods*, Wiley-Interscience, 1975.

Sokolnikoff, I. S., and Redheffer, R. M., *Mathematics of Physics and Modern Engineering*, McGraw Hill, 1966.

Wylie, C. R., *Advanced Engineering Mathematics*, McGraw Hill, 1975.

II. Structural Mechanics

Beer, F. S., and Johnston, E. R., Jr., *Vector Mechanics for Engineers*, McGraw Hill, 1972.

Fung, Y. C., *Foundations of Solid Mechanics*, Prentice Hall, 1965.

Langhaar, H. L., *Energy Methods in Applied Mechanics*, Wiley, 1962.

Martin, H. C., *Introduction to Matrix Methods of Structural Analysis*, McGraw Hill, 1966.

Pilkey, W. D., and Pilkey, O. H., *Mechanics of Solids*, Quantum, 1974.

Przemieniecki, J. S., *Theory of Matrix Structural Analysis*, McGraw Hill, 1968.

Tuma, J. J., *Theory and Problems of Structural Analysis*, Schaum-McGraw Hill, 1969.

Tuma, J. J., and Munshi, R. K., *Theory and Problems of Advanced Structural Analysis*, Schaum-McGraw Hill, 1971.

III. Finite Element Methods

Cheung, Y. K., and Yeo, M. F., *A Practical Introduction to Finite Element Analysis*, Pitman, 1979.

Cook, R. D., *Concepts and Applications of Finite Element Analysis*, Wiley, 1974.

Gallagher, R. H., *Finite Element Analysis—Fundamentals*, Prentice Hall, 1975.

Huebener, K. H., *The Finite Element Method for Engineers*, Wiley, 1975.

Martin, H. C., and Carey, C. F., *Introduction to Finite Element Analysis*, McGraw Hill, 1973.

Norrie, D. H., and deVries, G., *The Finite Element Method*, Academic Press, 1973.

Segerlind, L. J., *Applied Finite Element Analysis*, Wiley, 1976.

Spillers, W. R., *Automated Structural Analysis: An Introduction*, Pergamon, 1972.

Wachspress, E. L., *A Rational Finite Element Basis*, Academic Press, 1975.

Zienkiewicz, O. C., *The Finite Element Method in Structural and Continuum Mechanics*, McGraw Hill, 1980.

2

Trusses

2.1 INTRODUCTION

With the mathematical background now established, we are ready to discuss the finite-element method itself. We begin with an analysis of trusses. There are several reasons for starting here: First, from an historical point of view, the finite-element method was applied early with trusses—indeed, this was one of the first applications of the finite-element method; second, the physics or mechanics of trusses is simple and familiar; and finally, trusses provide an excellent means for introducing features of the finite-element method which are independent of particular applications.

Our basic approach is to develop the theory using simple, illustrative examples. (Later, we will take a more general, abstract approach.) Specifically, consider the two-dimensional truss shown in Figure 2.1. (Recall that a truss is a structural system of slender, "two-force" members or rods, joined with pins and loaded only

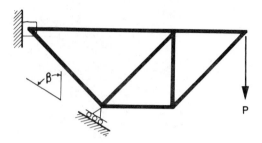

Figure 2.1   Truss Example

at the joints.*) A principal objective of this chapter is to find the displacements of the joints of this truss and the forces in its members when it is loaded and supported as shown. In the process of meeting this objective, we will develop an efficient method of analysis (the "finite-element-truss method") which is applicable, without conceptual modification, to any truss.

2.2 TRUSS ELEMENTS

To begin our discussion, consider a typical member of the truss such as is shown in Figure 2.2. Hooke's law states that if the tensile (or compressive) load in the member is f, then the elongation (or contraction) u is given by

$$u = f\ell/AE \tag{2.1}$$

where ℓ is the length of the member, A is its cross-sectional area, and E is Young's modulus of elasticity. Equation (2.1) may be

Figure 2.2   Typical Truss Member

---

*See, for example, Beer and Johnston, *Vector Mechanics for Engineers*, McGraw-Hill, New York, 1977, p. 214.

rewritten in the form

$$f = k u \qquad (2.2)$$

where k is defined as

$$k \stackrel{D}{=} AE/\ell \qquad (2.3)$$

To introduce some of the notation and some of the techniques useful in the finite-element method, consider the same truss member with the same loading (and, hence, the same deformation) but with the labeling and notation shown in Figures 2.3 and 2.4. (The "1" and "2" are labels designating the ends of the member.) By comparing Figures 2.2 and 2.3 and by equilibrium, we have

$$f_1 + f_2 = 0 \quad \text{or} \quad f_1 = -f_2 = -f \qquad (2.4)$$

Similarly, if we consider the elongation (or contraction) of the member in terms of the displacements of its ends as shown in Figure 2.4, we have

$$u = u_2 - u_1 \qquad (2.5)$$

(Note that if the member or bar of Figure 2.4 is unrestrained, rigid displacement can also occur. In either case, u is a measure of the relative displacement of the ends of the bar.)

Figure 2.3 Truss Member Force Notation

Figure 2.4 Truss Member Displacement Notation

By comparing Equations (2.2), (2.4), and (2.5), f_1 and f_2 may be expressed as

$$f_1 = ku_1 - ku_2 \tag{2.6}$$

and

$$f_2 = -ku_1 + ku_2 \tag{2.7}$$

or, by using matrix notation, we have

$$\{f\} = [k]\{u\} \tag{2.8}$$

where

$$\{f\} = \begin{Bmatrix} f_1 \\ f_2 \end{Bmatrix} , \quad \{u\} = \begin{Bmatrix} u_1 \\ u_2 \end{Bmatrix} \tag{2.9}$$

and

$$[k] = \begin{bmatrix} k & -k \\ -k & k \end{bmatrix} = k \begin{bmatrix} 1 & -1 \\ -1 & 1 \end{bmatrix} \tag{2.10}$$

The square matrix $[k]$ is called the "stiffness matrix" and in connection with this terminology the member or rod is called an "element" of the truss (hence, the name "finite-element" method).

2.3 ELEMENT STIFFNESS MATRIX

Prior to the use of the expression "finite-element method," many people referred to the same analysis as "matrix-structural analysis"— particularly when they were working with trusses. Moreover, the focus of attention in such analyses was (and is) the stiffness matrix such as described above. Therefore, to gain more familiarity with the procedures of the finite-element method, let us examine the stiffness matrix in more detail.

Regarding terminology, the stiffness matrix $[k]$ of Equation (2.10) is sometimes called a "local" or "element" stiffness matrix.

In the following sections we will see that it is convenient to also talk about a stiffness matrix for an entire truss, called a "global" stiffness matrix. The terms "local" or "element" and "global" are then used when it is necessary to distinguish between the two kinds of stiffness matrices.

Consider again the element of Figures 2.3 and 2.4 but this time let it be inclined to the horizontal as shown in Figure 2.5. If the forces f_1 and f_2 are resolved into horizontal and vertical components as shown in Figure 2.6, these components may be written

$$f_{1x} = f_1 \cos\theta$$
$$f_{1y} = f_1 \sin\theta$$

(2.11)

and

$$f_{2x} = f_2 \cos\theta$$
$$f_{2y} = f_2 \sin\theta$$

(2.12)

By multiplying the first of Equations (2.11) by $\cos\theta$ and the second by $\sin\theta$ and adding, we can solve for f_1 as

$$f_1 = f_{1x} \cos\theta + f_{1y} \sin\theta$$

(2.13)

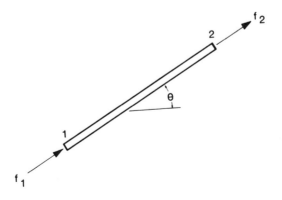

Figure 2.5   Inclined Truss Element

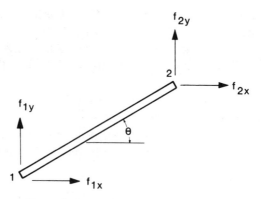

Figure 2.6 Inclined Truss Element with
Equivalent Horizontal and Vertical Loading

Similarly, from Equations (2.12), we obtain f_2 as

$$f_2 = f_{2x} \cos\theta + f_{2y} \sin\theta \tag{2.14}$$

An analogous analysis could be made for the displacements.
However, we must be careful to recognize that the displacement
vectors u_1 and u_2 at ends 1 and 2 are *not* necessarily parallel to the
member. That is, even though the forces at ends 1 and 2 are parallel
to the member (since it is a "two-force" member), the displacements
need not be parallel to the member because of the possibility of
"rigid displacement" of the truss member due to the overall deformation of the truss. Hence, to account for this let u be the projection
of the relative displacements along the member. That is, let u be

$$u = (u_2 - u_1) \cdot n \tag{2.15}$$

where n is a unit vector parallel to the member as shown in Figure
2.7. Also shown in Figure 2.7 are the horizontal and vertical unit
vectors n_x and n_y. In terms of n_x and n_y, u_1, u_2, and n are

$$u_1 = u_{1x} n_x + u_{1y} n_y \tag{2.16}$$

$$u_2 = u_{2x} n_x + u_{2y} n_y \tag{2.17}$$

and

$$n = \cos\theta n_x + \sin\theta n_y \tag{2.18}$$

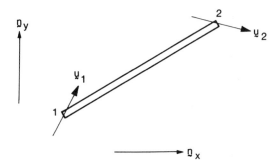

Figure 2.7   Truss Member and Associated
Displacement and Unit Vectors

By substituting into Equation (2.15) the relative displacement along the member, u, is

$$u = u_{2x} \cos\theta + u_{2y} \sin\theta - u_{1x} \cos\theta - u_{1y} \sin\theta \qquad (2.19)$$

Since from Equations (2.2) and (2.4) $f_1 = -f_2 = -f = -ku$, f_1 and f_2 become

$$f_1 = ku_{1x} \cos\theta + ku_{1y} \sin\theta - ku_{2x} \cos\theta - ku_{2y} \sin\theta \qquad (2.20)$$

and

$$f_2 = -k u_{1x} \cos\theta - k u_{1y} \sin\theta + k u_{2x} \cos\theta + k u_{2y} \sin\theta \qquad (2.21)$$

Finally, substitution of Equations (2.20) and (2.21) into (2.11) and (2.12) leads to

$$\begin{aligned}
f_{1x} &= (k \cos^2\theta) u_{1x} + (k \sin\theta \, \cos\theta) u_{1y} + (-k \cos^2\theta) u_{2x} \\
&\quad + (-k \sin\theta \, \cos\theta) u_{2y}
\end{aligned}$$

$$\begin{aligned}
f_{1y} &= (k \sin\theta \, \cos\theta) u_{1x} + (k \sin^2\theta) u_{1y} + (-k \sin\theta \, \cos\theta) u_{2x} \\
&\quad + (-k \sin^2\theta) u_{2y} \qquad (2.22)
\end{aligned}$$

$$\begin{aligned}
f_{2x} &= (-k \cos^2\theta) u_{1x} + (-k \sin\theta \, \cos\theta) u_{1y} + (k \cos^2\theta) u_{2x} \\
&\quad + (k \sin\theta \, \cos\theta) u_{2y}
\end{aligned}$$

$$\begin{aligned}
f_{2y} &= (-k \sin\theta \, \cos\theta) u_{1x} + (-k \sin^2\theta) u_{1y} + (k \sin\theta \, \cos\theta) u_{2x} \\
&\quad + (k \sin^2\theta) u_{2y}
\end{aligned}$$

These equations may be compactly written in matrix form as

$$\{f\} = [k]\,\{u\} \tag{2.23}$$

where now [see Equations (2.8), (2.9) and (2.10)] $\{f\}$ and $\{u\}$ are the column vectors

$$\{f\} = \begin{Bmatrix} f_{1x} \\ f_{1y} \\ f_{2x} \\ f_{2y} \end{Bmatrix} \qquad\qquad \{u\} = \begin{Bmatrix} u_{1x} \\ u_{1y} \\ u_{2x} \\ u_{2y} \end{Bmatrix} \tag{2.24}$$

and the stiffness matrix $[k]$ is now

$$[k] = k \begin{bmatrix} c^2 & sc & -c^2 & -sc \\ sc & s^2 & -sc & -s^2 \\ -c^2 & -sc & c^2 & sc \\ -sc & -s^2 & sc & s^2 \end{bmatrix} \tag{2.25}$$

where s and c are abbreviations for $\sin\theta$ and $\cos\theta$, respectively.

At this point, it might be helpful to make several remarks: First, the stiffness matrix is symmetrical; that is, the respective rows and columns are interchangeable. Second, it is singular; that is, its determinant is zero. Third, the sum of the elements in a given row or column is zero. Fourth, the formal similarity of Equations (2.2), (2.8), and (2.9) is obvious. Indeed, Equation (2.22) could be viewed as a generalization of Equation (2.2) or (2.8). Finally, if θ is zero in Equation (2.25), $[k]$ becomes

$$[k] = k \begin{bmatrix} 1 & 0 & -1 & 0 \\ 0 & 0 & 0 & 0 \\ -1 & 0 & 1 & 0 \\ 0 & 0 & 0 & 0 \end{bmatrix} \tag{2.26}$$

2.4 DEVELOPMENT OF THE STIFFNESS MATRIX USING TRANSFORMATION MATRICES

The stiffness matrices of Equations (2.25) and (2.26) can be developed from the stiffness matrix of Equation (2.10) by expansion and transformation as follows: Consider the truss element to be loaded as shown in Figure 2.3. Next, introduce forces perpendicular to the element, as shown in Figure 2.8. (We will shortly assign zero magnitudes to these forces.) Similarly, introduce displacements perpendicular to the element as shown in Figure 2.9. (Regarding notation, the primes on the force and displacement symbols of Figures 2.8 and 2.9 are simply used to distinguish them from the forces and displacements of inclined truss elements, as in Figure 2.6.)

By assuming a linear relationship between the forces and the displacements (as in Equations (2.2), (2.8), and (2.22), the forces of Figure 2.7 may be related to the displacements of Figure 2.8 as follows:

$$
\begin{bmatrix} f'_{1x} \\ f'_{1y} \\ f'_{2x} \\ f'_{2y} \end{bmatrix} = \begin{bmatrix} k'_{11} & k'_{12} & k'_{13} & k'_{14} \\ k'_{21} & k'_{22} & k'_{23} & k'_{24} \\ k'_{31} & k'_{32} & k'_{33} & k'_{34} \\ k'_{41} & k'_{42} & k'_{43} & k'_{44} \end{bmatrix} \begin{bmatrix} u'_{1x} \\ u'_{1y} \\ u'_{2x} \\ u'_{2y} \end{bmatrix} \tag{2.27}
$$

Figure 2.8   Truss Element with Perpendicular (Zero Magnitude) Forces

Figure 2.9   Displacement Components of the Truss Element

where the elements of the stiffness matrix k'_{ij} $(i, j = 1, \ldots, 4)$ are to be determined. We can easily determine them, however, by observing that we can make the magnitudes of f'_{1y} and f'_{2y} be zero by setting the elements of the second and fourth rows to zero. Also, from physical reasoning, we can argue that the displacements u'_{1y} and u'_{2y} should not affect the magnitudes of f'_{1x} and f'_{2x} (for otherwise the fundamental relationship of Equation (2.2) would be violated). We can prevent u'_{1y} and u'_{2y} from affecting the magnitudes of f'_{1x} and f'_{2x} by making the elements in the second and fourth columns of the stiffness matrix be zero. Equation (2.27) then becomes

$$
\begin{bmatrix} f'_{1x} \\ f'_{1y} \\ f'_{2x} \\ f'_{2y} \end{bmatrix} = \begin{bmatrix} k'_{11} & 0 & k'_{13} & 0 \\ 0 & 0 & 0 & 0 \\ k'_{31} & 0 & k'_{33} & 0 \\ 0 & 0 & 0 & 0 \end{bmatrix} \begin{bmatrix} u'_{1x} \\ u'_{1y} \\ u'_{2x} \\ u'_{2y} \end{bmatrix} \tag{2.28}
$$

The remaining unknown elements may be determined by recognizing that Equation (2.28) is simply an expanded version of Equation (2.8) (with $f'_{1x}, f'_{2x}, u'_{1x}$, and u'_{2x} identified with f_1, f_2, u_1, and u_2, respectively). Hence, $k'_{11}, k'_{13}, k'_{31}$, and k'_{33} have the values $k, -k, k$, and $-k$ respectively and Equation (2.28) finally becomes

$$
\begin{bmatrix} f'_{1x} \\ f'_{1y} \\ f'_{2x} \\ f'_{2y} \end{bmatrix} = k \begin{bmatrix} 1 & 0 & -1 & 0 \\ 0 & 0 & 0 & 0 \\ -1 & 0 & 1 & 0 \\ 0 & 0 & 0 & 0 \end{bmatrix} \begin{bmatrix} u'_{1x} \\ u'_{1y} \\ u'_{2x} \\ u'_{2y} \end{bmatrix} \tag{2.29}
$$

We have thus developed the stiffness matrix of Equation (2.26) by basically expanding, that is, "filling in" zero rows and columns, in the stiffness matrix of Equation (2.10). This technique will also be useful later when we are developing global stiffness matrices for the entire truss.

We can develop the general stiffness matrix of Equation (2.25) from the stiffness matrix of Equation (2.29) and hence, from that of Equation (2.10) by using transformation matrices as follows:

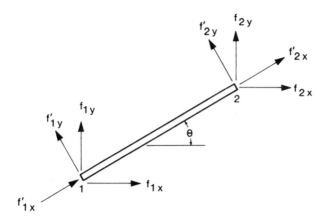

Figure 2.10   Superimposition of the Forces of Figures 2.6 and 2.7

Consider a comparison of the forces in Figure 2.6 and 2.8. If these figures are superimposed, as in Figure 2.10, it is seen that

$$f'_{1x} = f_{1x} \cos\theta + f_{1y} \sin\theta \tag{2.30}$$

and

$$f'_{1y} = -f_{1x} \sin\theta + f_{1y} \cos\theta \tag{2.31}$$

In matrix form, these equations are

$$\begin{Bmatrix} f'_{1x} \\ f'_{1y} \end{Bmatrix} = \begin{bmatrix} c & s \\ -s & c \end{bmatrix} \begin{Bmatrix} f_{1x} \\ f_{1y} \end{Bmatrix} = [t] \begin{Bmatrix} f_{1x} \\ f_{1y} \end{Bmatrix} \tag{2.32}$$

where as before s and c are abbreviations for $\sin\theta$ and $\cos\theta$ and where the transformation matrix $[t]$ is defined by the equation itself. Note that $[t]$ is orthogonal, that is, its inverse is equal to its transpose.

$$[t]^{-1} = [t]^T \tag{2.33}$$

Similarly, for the other end of the element, we have

$$\begin{Bmatrix} f'_{2x} \\ f'_{2y} \end{Bmatrix} = [t] \begin{Bmatrix} f_{2x} \\ f_{2y} \end{Bmatrix} \tag{2.34}$$

Hence, using partitioned matrices, we can write

$$
\begin{Bmatrix} f'_{1x} \\ f'_{1y} \\ -- \\ f'_{2x} \\ f'_{2y} \end{Bmatrix}
=
\begin{bmatrix} t & | & 0 \\ --L-- \\ 0 & | & t \end{bmatrix}
\begin{Bmatrix} f_{1x} \\ f_{1y} \\ -- \\ f_{2x} \\ f_{2y} \end{Bmatrix}
= [T]
\begin{Bmatrix} f_{1x} \\ f_{1y} \\ -- \\ f_{2x} \\ f_{2y} \end{Bmatrix}
\tag{2.35}
$$

where the 4 X 4 transformation matrix $[T]$ is defined by the equation itself. Note that like $[t]$, $[T]$ is also orthogonal, that is,

$$
[T]^{-1} = [T]^{T} \tag{2.36}
$$

Similarly, it is seen by direct analogy that

$$
\begin{Bmatrix} u'_{1x} \\ u'_{1y} \\ u'_{2x} \\ u'_{2y} \end{Bmatrix}
= [T]
\begin{Bmatrix} u_{1x} \\ u_{1y} \\ u_{2x} \\ u_{2y} \end{Bmatrix}
\tag{2.37}
$$

Equations (2.34) and (2.38) may be written compactly as

$$
\{f'\} = [T]\{f\} \tag{2.38}
$$

and

$$
\{u'\} = [T]\{u\} \tag{2.39}
$$

where the notation is self-explanatory. Recall, however, from Equation (2.29) that

$$
\{f'\} = [k']\{u'\} \tag{2.40}
$$

where $[k']$ is the stiffness matrix of Equation (2.26). By substituting Equations (2.38) and (2.39) into (2.40) we have

$$[T] \{f\} = [k'] [T] \{u\} \tag{2.41}$$

or

$$\{f\} = [T]^T [k'] [T]\{u\} = [k]\{u\} \tag{2.42}$$

Therefore,

$$[k] = [T]^T [k'] [T] = \begin{bmatrix} c & -s & & \\ s & c & & 0 \\ \hline & & c & -s \\ 0 & & s & c \end{bmatrix} \begin{bmatrix} k & 0 & -k & 0 \\ 0 & 0 & 0 & 0 \\ \hline -k & 0 & k & 0 \\ 0 & 0 & 0 & 0 \end{bmatrix} \begin{bmatrix} c & s & & \\ -s & c & & 0 \\ \hline & & c & s \\ 0 & & -s & c \end{bmatrix}$$

$$\tag{2.43}$$

By expanding the indicated matrix product, we obtain exactly the same stiffness matrix as in Equation (2.25).

2.5 GLOBAL STIFFNESS AND ASSEMBLY PROCEDURES

Recall that our basic objective is to develop an efficient procedure for finding the joint displacements and the element forces for any given loading and support conditions. We are now ready to take a major step in meeting this objective. Indeed, in this section we will consider a way to expand the foregoing matrix analysis of a single element to an analysis of an entire truss such as is shown in Figure 2.1. The procedure is simply to construct or assemble the truss via its elements: that is, force, stiffness, and displacement arrays for the entire truss will be obtained by expansion and superposition of these arrays on an element-by-element basis. This process of "putting it all together" is called the "assembly procedure."

The procedure is really quite straightforward. Indeed, its essential features can quickly be grasped by observing an example assembly. Therefore, consider the simple, three-element truss shown in Figure 2.11 where force P is given. The objective is to determine (i) the resulting forces in each of the three elements; (ii) the unknown joint displacements; and (iii) the unknown reaction forces.

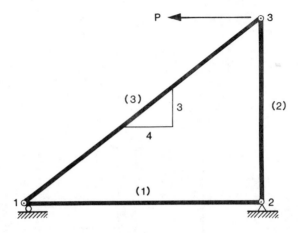

Figure 2.11   Simple Truss Example

To meet this objective, it is helpful to temporarily consider the more general three-element truss shown in Figure 2.12. Here, we assume that there are forces with horizontal and vertical components applied at each joint, as shown, and that these forces are such that the truss as a unit is in static equilibrium. The displacements at the joints are as shown in Figure 2.13. Our objective of finding the unknown forces and displacements will be essentially satisfied if we can find a stiffness matrix $[k]$ such that

$$\{F\} = [K]\{U\} \tag{2.44}$$

where $\{F\}$ is the force array

$$\{F\} = \begin{bmatrix} F_{1x} \\ F_{1y} \\ F_{2x} \\ F_{2y} \\ F_{3x} \\ F_{3y} \end{bmatrix} = \begin{bmatrix} F_1 \\ F_2 \\ F_3 \\ F_4 \\ F_5 \\ F_6 \end{bmatrix} \tag{2.45}$$

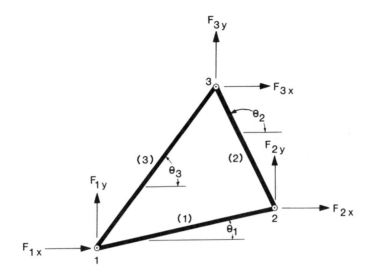

Figure 2.12   General Three-Element Truss and Forces

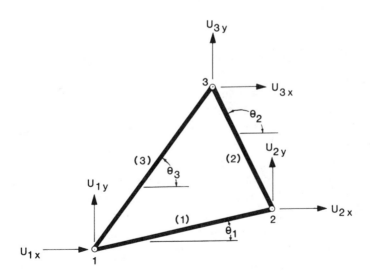

Figure 2.13   General Three-Element Truss Displacements

$\{U\}$ is the displacement array

$$\{U\} = \begin{bmatrix} U_{1x} \\ U_{1y} \\ U_{2x} \\ U_{2y} \\ U_{3x} \\ U_{3y} \end{bmatrix} = \begin{bmatrix} U_1 \\ U_2 \\ U_3 \\ U_4 \\ U_5 \\ U_6 \end{bmatrix} \qquad (2.46)$$

and $[K]$ is the 6 × 6 matrix

$$[K] = \begin{bmatrix} K_{11} & K_{12} & K_{13} & K_{14} & K_{15} & K_{16} \\ K_{21} & K_{22} & K_{23} & K_{24} & K_{25} & K_{26} \\ K_{31} & K_{32} & K_{33} & K_{34} & K_{35} & K_{36} \\ K_{41} & K_{42} & K_{43} & K_{44} & K_{45} & K_{46} \\ K_{51} & K_{52} & K_{53} & K_{54} & K_{55} & K_{56} \\ K_{61} & K_{62} & K_{63} & K_{64} & K_{65} & K_{66} \end{bmatrix} \qquad (2.47)$$

Now, since notation is often the most confusing aspect of finite-element analysis, it might be helpful at this point to make a few remarks about our nomenclature and notation convention. We will try to maintain this convention throughout the text. (a) The truss joints are labeled by simple indexing or numbering and the truss elements are labeled by an indexing or numbering in parentheses. (b) Capital or upper class letters (majuscules) are used for global (that is, entire truss) quantities and lower case letters (miniscules) are used for local (that is, element) quantities. (c) The global force and displacement arrays are sometimes written as single-indexed, ordered arrays as in the right side of Equations (2.45) and (2.46). (This is simply an aid for computational purposes.)

In addition to this, note that the truss of Figure 2.12 has three elements and that the local or element force-displacement relations [Equation (2.22)] are of the same form for all three elements. Therefore, to distinguish between the elements, we will place an

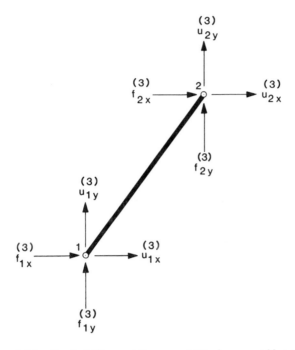

Figure 2.14   Typical Element Force and Displacement Notation

element label in superscripted parentheses on the force, displace-ment, and stiffness arrays and matrices. For example, the forces and displacements for element (3) would be shown in Figure 2.14. Note also that the subscripts "1" and "2" are still retained locally to designate the ends of the element even though the joints have, in general, different global numbers. We will generally use the con-vention of assigning the local "1" to the lowered numbered global joint and thus the local "2" will be assigned to the higher numbered global joint.

To return to the analysis, consider the free-body diagram in Figure 2.15 of joint 1 of the truss of Figure 2.12. Note that the free-body diagram includes all the forces acting on the joint, that is, both global (external) and local (internal) forces. Note also that because of the law of action and reaction,* the directions of the element

*See, for example, Kane, *Analytical Elements of Mechanics*, Academic Press, New York, 1959, Vol. 1, p. 150.

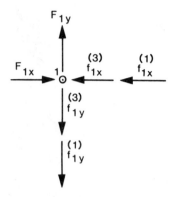

Figure 2.15   Free-Body Diagram of Joint 1 of the Truss of Figure 2.12

forces are opposite to the directions of those shown in Figure 2.14. Equilibrium considerations lead to the equations

$$F_{1x} = f_{1x}^{(1)} + f_{1x}^{(3)}$$
$$F_{1y} = f_{1y}^{(1)} + f_{1y}^{(3)} \tag{2.48}$$

Similarly, from free-body diagrams of joints 2 and 3, we obtain

$$F_{2x} = f_{2x}^{(1)} + f_{1x}^{(2)}$$
$$F_{2y} = f_{2y}^{(1)} + f_{1y}^{(2)} \tag{2.49}$$

and

$$F_{3x} = f_{2x}^{(2)} + f_{2x}^{(3)}$$
$$F_{3y} = f_{2y}^{(2)} + f_{2y}^{(3)} \tag{2.50}$$

Equations (2.48), (2.49), and (2.50) may be written in matrix form as

$$
\begin{bmatrix} F_{1x} \\ F_{1y} \\ F_{2x} \\ F_{2y} \\ F_{3x} \\ F_{3y} \end{bmatrix}
=
\begin{bmatrix} f_{1x}^{(1)} \\ f_{1y}^{(1)} \\ f_{2x}^{(1)} \\ f_{2y}^{(1)} \\ 0 \\ 0 \end{bmatrix}
+
\begin{bmatrix} 0 \\ 0 \\ f_{1x}^{(2)} \\ f_{1y}^{(2)} \\ f_{2x}^{(2)} \\ f_{2y}^{(2)} \end{bmatrix}
+
\begin{bmatrix} f_{1x}^{(3)} \\ f_{1y}^{(3)} \\ 0 \\ 0 \\ f_{2y}^{(3)} \\ f_{2y}^{(3)} \end{bmatrix}
\tag{2.51}
$$

However by using Equations (2.23) and (2.25), and by expanding the size of the arrays on those equations by inserting rows and columns with zero elements, as in the foregoing secton, the arrays on the right side of Equation (2.51) may be written

$$
\begin{bmatrix} f_{1x}^{(1)} \\ f_{1y}^{(1)} \\ f_{2x}^{(1)} \\ f_{2y}^{(1)} \\ 0 \\ 0 \end{bmatrix} = k_1 \begin{bmatrix} c_1^2 & s_1 c_1 & -c_1^2 & -s_1 c_1 & 0 & 0 \\ s_1 c_1 & s_1^2 & -s_1 c_1 & -s_1^2 & 0 & 0 \\ -c_1^2 & -s_1 c_1 & c_1^2 & s_1 c_1 & 0 & 0 \\ -s_1 c_1 & -s_1^2 & s_1 c_1 & s_1^2 & 0 & 0 \\ 0 & 0 & 0 & 0 & 0 & 0 \\ 0 & 0 & 0 & 0 & 0 & 0 \end{bmatrix} \begin{bmatrix} u_{1x}^{(1)} \\ u_{1y}^{(1)} \\ u_{2x}^{(1)} \\ u_{2y}^{(1)} \\ 0 \\ 0 \end{bmatrix}
$$

$$(2.52)$$

$$
\begin{bmatrix} 0 \\ 0 \\ f_{1x}^{(2)} \\ f_{1y}^{(2)} \\ f_{2x}^{(2)} \\ f_{2y}^{(2)} \end{bmatrix} = k_2 \begin{bmatrix} 0 & 0 & 0 & 0 & 0 & 0 \\ 0 & 0 & 0 & 0 & 0 & 0 \\ 0 & 0 & c_2^2 & s_2 c_2 & -c_2^2 & -s_2 c_2 \\ 0 & 0 & s_2 c_2 & s_2^2 & -s_2 c_2 & -s_2^2 \\ 0 & 0 & -c_2^2 & -s_2 c_2 & c_2^2 & s_2 c_2 \\ 0 & 0 & -s_2 c_2 & -s_2^2 & s_2 c_2 & s_2^2 \end{bmatrix} \begin{bmatrix} 0 \\ 0 \\ u_{1x}^{(2)} \\ u_{1y}^{(2)} \\ u_{2x}^{(2)} \\ u_{2y}^{(2)} \end{bmatrix}
$$

$$(2.53)$$

and

$$
\begin{bmatrix} f_{1x}^{(3)} \\ f_{1y}^{(3)} \\ 0 \\ 0 \\ f_{2x}^{(3)} \\ f_{2y}^{(3)} \end{bmatrix} = k_3 \begin{bmatrix} c_3^2 & s_3 c_3 & 0 & 0 & -c_3^2 & -s_3 c_3 \\ s_3 c_3 & s_3^2 & 0 & 0 & -s_3 c_3 & -s_3^2 \\ 0 & 0 & 0 & 0 & 0 & 0 \\ 0 & 0 & 0 & 0 & 0 & 0 \\ -c_3^2 & -s_3 c_3 & 0 & 0 & c_3^2 & s_3 c_3 \\ -s_3 c_3 & -s_3^2 & 0 & 0 & s_3 c_3 & s_3^2 \end{bmatrix} \begin{bmatrix} u_{1x}^{(3)} \\ u_{1y}^{(3)} \\ 0 \\ 0 \\ u_{2x}^{(3)} \\ u_{2y}^{(3)} \end{bmatrix}
$$

$$(2.54)$$

where the subscripts on s_i and c_i (i = 1, 2, 3) refer to the inclination angles θ_1, θ_2, or θ_3 of the truss elements (for example, $c_2 = \cos\theta_2$), and where the subscript on k refers to the respective truss elements.

The local displacements in Equations (2.52), (2.53), and (2.54) may be identified with the global displacements U_i $(i = 1, \ldots, 6)$ as follows

$$
\begin{aligned}
u_{1x}^{(1)} &= u_{1x}^{(3)} = U_{1x} = U_1 \\
u_{1y}^{(1)} &= u_{1y}^{(3)} = U_{1y} = U_2 \\
u_{2x}^{(1)} &= u_{1x}^{(2)} = U_{2x} = U_3 \\
u_{2y}^{(1)} &= u_{1y}^{(2)} = U_{2y} = U_4 \\
u_{2x}^{(2)} &= u_{2x}^{(3)} = U_{3x} = U_5 \\
u_{2y}^{(2)} &= u_{2y}^{(3)} = U_{3y} = U_6
\end{aligned}
\tag{2.55}
$$

Hence, by substituting Equations (2.52), (2.53), (2.54), and (2.55) into Equation (2.51) and by collecting terms, we obtain an expression of the form

$$
\{F\} = ([K^{(1)}] + [K^{(2)}] + [K^{(3)}]) \{U\} = [K] \{U\} \tag{2.56}
$$

where $[K^{(1)}]$, $[K^{(2)}]$, and $[K^{(3)}]$ are the expanded element stiffness matrices of Equations (2.52), (2.53), and (2.54), and the superscripted index refers to the respective element.

Finally, therefore, from Equation (2.56) the desired global stiffness matrix $[K]$ is

$$
[K] = [K^{(1)}] + [K^{(2)}] + [K^{(3)}]
$$

$$
\begin{bmatrix}
(k_1 c_1^2 + k_3 c_3^2) & (k_1 s_1 c_1 + k_3 s_3 c_3) & (-k_1 c_1^2) & (-k_1 s_1 c_1) & (-k_3 c_3^2) & (-k_3 s_3 c_3) \\
(k_1 s_1 c_1 + k_3 s_3 c_3) & (k_1 s_1^2 + k_3 s_3^2) & (-k_1 s_1 c_1) & (-k_1 s_2^2) & (-k_3 s_3 c_3) & (-k_3 s_3^2) \\
(-k_1 c_1^2) & (-k_1 s_1 c_1) & (k_1 c_1^2 + k_2 c_2^2) & (k_1 s_1 c_1 + k_2 s_2 c_2) & (-k_2 c_2^2) & (-k_2 s_2 c_2) \\
(-k_1 s_1 c_1) & (-k_1 s_1^2) & (k_1 s_1 c_1 + k_2 s_2 c_2) & (k_1 s_1^2 + k_2 s_2^2) & (-k_2 s_2 c_2) & (-k_2 s_2^2) \\
(-k_3 c_2^2) & (-k_3 s_3 c_3) & (-k_2 c_2^2) & (-k_2 s_2 c_2) & (k_2 c_2^2 + k_3 c_3^2) & (k_2 s_2 c_2 + k_3 s_3 c_3) \\
(-k_3 s_3 c_3) & (-k_3 s_3^2) & (-k_2 s_2 c_2) & (-k_2 s_2^2) & (k_2 s_2 c_2 + k_3 s_3 c_3) & (k_2 s_2^2 + k_3 s_3^2)
\end{bmatrix}
$$

$$
\tag{2.57}
$$

Before we return to the simple truss of Figure 2.11, it is interesting to observe that like the local stiffness matrices, the global stiffness matrix $[K]$ also has the properties that (1) it is symmetric; (b) the sum of the elements in any row or column is zero; and (c) it is singular, that is, its determinant is zero.

Now, to use Equations (2.56) and (2.57) to find the unknown forces and displacements of our simple example truss of Figure 2.11, we note that for that truss

$$\theta_1 = 0, \qquad \theta_2 = \pi/2, \qquad \theta_3 = \sin^{-1} 3/5,$$

$$\tag{2.58}$$

$$k^{(1)} = k_1, \qquad k_1^{(2)} = k_2, \qquad k_1^{(3)} = k_3$$

Also from Figure 2.11 we can easily deduce the following known forces and displacements

$$F_{1x} = F_1 = 0, \quad F_{3x} = F_5 = -P, \quad F_{3y} = F_6 = 0$$

and
$$\tag{2.59}$$

$$U_{1y} = U_2 = 0, \quad U_{2x} = U_3 = 0, \quad U_{2y} = U_4 = 0$$

Hence, Equation (2.56) becomes

$$
\begin{bmatrix} 0 \\ F_2 \\ F_3 \\ F_4 \\ -P \\ 0 \end{bmatrix} =
\begin{bmatrix}
k_1 + k_3\left(\tfrac{4}{5}\right)^2 & k_3\left(\tfrac{3}{5}\right)\left(\tfrac{4}{5}\right) & -k_1 & 0 & -k_3\left(\tfrac{4}{5}\right)^2 & -k_3\left(\tfrac{3}{5}\right)\left(\tfrac{4}{5}\right) \\
k_3\left(\tfrac{3}{5}\right)\left(\tfrac{4}{5}\right) & k_3\left(\tfrac{3}{5}\right)^2 & 0 & 0 & -k_3\left(\tfrac{3}{5}\right)\left(\tfrac{4}{5}\right) & -k_3\left(\tfrac{3}{5}\right)^2 \\
-k_1 & 0 & k_1 & 0 & 0 & 0 \\
0 & 0 & 0 & k_2 & 0 & -k_2 \\
-k_3\left(\tfrac{4}{5}\right)^2 & -k_3\left(\tfrac{3}{5}\right)\left(\tfrac{4}{5}\right) & 0 & 0 & k_3\left(\tfrac{4}{5}\right)^2 & k_3\left(\tfrac{3}{5}\right)\left(\tfrac{4}{5}\right) \\
-k_3\left(\tfrac{3}{5}\right)\left(\tfrac{4}{5}\right) & -k_3\left(\tfrac{3}{5}\right)^2 & 0 & -k_2 & k_3\left(\tfrac{3}{5}\right)\left(\tfrac{4}{5}\right) & k_2 + k_3\left(\tfrac{3}{5}\right)^2
\end{bmatrix}
\begin{bmatrix} U_1 \\ 0 \\ 0 \\ 0 \\ U_5 \\ U_6 \end{bmatrix}
$$

$$\tag{2.60}$$

or in single equation form

$$0 = [k_1 + k_3(\tfrac{4}{5})^2] U_1 - k_3(\tfrac{4}{5})^2 U_5 - k_3(\tfrac{3}{5})(\tfrac{4}{5}) U_6$$

$$F_2 = k_3(\tfrac{3}{5})(\tfrac{4}{5}) U_1 - k_3(\tfrac{3}{5})(\tfrac{4}{5}) U_5 - k_3(\tfrac{3}{5})^2 U_6$$

$$F_3 = -k_1 U_1$$

$$F_4 = -k_2 U_6 \tag{2.61}$$

$$-P = -k_3(\tfrac{4}{5})^2 U_1 + k_3(\tfrac{4}{5})^2 U_5 + k_3(\tfrac{3}{5})(\tfrac{4}{5}) U_6$$

$$0 = -k_3(\tfrac{3}{5})(\tfrac{4}{5}) U_1 + k_3(\tfrac{3}{5})(\tfrac{4}{5}) U_5 + [k_2 + k_3(\tfrac{3}{5})^2] U_6$$

Solving for U_1, U_2, U_3, F_2, F_3, and F_4, we obtain

$$U_1 = -P/k_1$$

$$U_5 = -P[(1/k_1) + (3/4)^2(1/k_2) + (5/4)^2(1/k_3)]$$

$$U_6 = (3/4)P/k_2$$

$$F_2 = (3/4)P \tag{2.62}$$

$$F_3 = P$$

$$F_4 = -(3/4)P$$

By knowing the forces and displacements at the joints, the methods of the elementary structural analysis can readily be used to determine the forces in the individual members of the truss.

2.6 ASSEMBLY OF LARGE TRUSSES

Although the assembly procedure outlined in the foregoing section is conceptually very simple, it can also be quite tedious—as we have seen—even for the small truss of Figure 2.11. For larger trusses such as our fundamental example truss of Figure 2.1, and reproduced

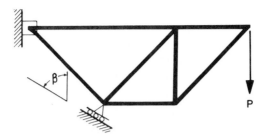

Figure 2.16   Truss Example

in Figure 2.16, the procedure can be very laborious unless it is automated or symbolically organized. Therefore, the objective of this section is to formally organize the assembly procedure so that it may readily be applied with large trusses. We will develop and illustrate this "organized" assembly procedure with the truss of Figure 2.17.

Figure 2.17 shows a typical labelling of the elements and joints of the example truss. The relations between the global and local displacements may be summarized by the following equations:

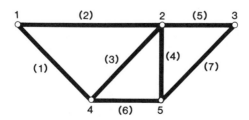

Figure 2.17   Truss Element and Joint Labels

$$1 \Big\langle \begin{array}{l} U_1 = U_{1x} = u_{1x}^{(1)} = u_{1x}^{(2)} \\[2mm] U_2 = U_{1y} = u_{1y}^{(1)} = u_{1y}^{(2)} \end{array}$$

$$2 \Big\langle \begin{array}{l} U_3 = U_{2x} = \quad = u_{2x}^{(2)} = u_{1x}^{(3)} = u_{1x}^{(4)} = u_{1x}^{(5)} \\[2mm] U_4 = U_{2y} = \quad = u_{2y}^{(2)} = u_{1y}^{(3)} = u_{1y}^{(4)} = u_{1y}^{(5)} \end{array}$$

$$3 \Big\langle \begin{array}{l} U_5 = U_{3x} = \qquad\qquad\qquad = u_{2x}^{(5)} \qquad = u_{1x}^{(7)} \\[2mm] U_6 = U_{3y} = \qquad\qquad\qquad = u_{2y}^{(5)} \qquad = u_{1y}^{(7)} \end{array}$$

$$4 \Big\langle \begin{array}{l} U_7 = U_{4x} = u_{2x}^{(1)} \qquad = u_{2x}^{(3)} \qquad\qquad = u_{1x}^{(6)} \\[2mm] U_8 = U_{4y} = u_{2y}^{(1)} \qquad = u_{2y}^{(3)} \qquad\qquad = u_{1y}^{(6)} \end{array}$$

$$5 \Big\langle \begin{array}{l} U_9 = U_{5x} = \qquad\qquad = u_{2x}^{(4)} \qquad = u_{2x}^{(6)} = u_{2x}^{(7)} \\[2mm] U_{10} = U_{5y} = \qquad\qquad = u_{2y}^{(4)} \qquad = u_{2y}^{(6)} = u_{2y}^{(7)} \end{array}$$

$$(2.63)$$

Notice that it is natural to group these equations and displacements in pairs and to identify these equation pairs with the truss joints. These equations also show that for each truss element there is a relationship between the local and global displacement and that this relationship can be written in the form

$$\{u^{(e)}\} = [\Lambda^{(e)}] \{U\} \tag{2.64}$$

where $[\Lambda^{(e)}]$ is a 4 × 10 array of one's and zero's arranged for each element to be consistent with the equations. For example, for truss element (1), $[\Lambda^{(1)}]$ is

$$[\Lambda^{(1)}] = \begin{bmatrix} 1 & 0 & 0 & 0 & 0 & 0 & 0 & 0 & 0 & 0 \\ 0 & 1 & 0 & 0 & 0 & 0 & 0 & 0 & 0 & 0 \\ 0 & 0 & 0 & 0 & 0 & 0 & 1 & 0 & 0 & 0 \\ 0 & 0 & 0 & 0 & 0 & 0 & 0 & 1 & 0 & 0 \end{bmatrix} \tag{2.65}$$

The $[\overset{(e)}{\Lambda}]$ arrays are called "incidence matrices." They may be systematically developed as follows: (a) Let all the elements of the $[\overset{(e)}{\Lambda}]$ matrix initially be zero; (b) group the columns of the matrix in pairs and associate successive pairs with the truss joints, as in Table 2.1; (c) using Figure 2.17 construct a table identifying the "1" and "2" ends of the truss elements with the joint numbers, as in Table 2.2; and (d) construct a third table which identifies for each truss element the row and column of the nonzero (that is, "one's") entries to be inserted in the $[\overset{(e)}{\Lambda}]$ matrix, as in Table 2.3.

Table 2.1   Truss-Joint/Matrix-Column Matching

Joint Number	Column Numbers	
1	1	2
2	3	4
3	5	6
4	7	8
5	9	10

Table 2.2   Truss-Elements-Ends/Joint Number Identification

Truss Element e	Joint Numbers N_1	N_2
1	1	4
2	1	2
3	2	4
4	2	5
5	2	3
6	4	5
7	3	5

Table 2.3   Nonzero $[\overset{(e)}{\Lambda}]$ Row-Column Entries

Row	Truss Element e						
	1	2	3	4	5	6	7
1	1	1	3	3	3	7	5
2	2	2	4	4	4	8	6
3	7	3	7	9	5	9	9
4	8	4	8	10	6	10	10

Column

Table 2.3 may be constructed from Tables 2.1 and 2.2 by matching the joint numbers with the column numbers. Specifically: (a) For each truss element, determine the corresponding joint number from Table 2.2. (b) For these joint numbers, obtain, from Table 2.1, the column numbers of the nonzero entries in the incidence matrix. (c) List these column numbers successively under the element numbers in the columns of Table 2.3. (Table 2.3 will be useful later in the sequel in constructing the global stiffness matrix.) Regarding notation, the ends of the elements are sometimes called "nodes"—hence, the "N_1" and "N_2" in Table 2.2. This notation will also be used in the sequel.

Using Table 2.3, the $[\overset{(e)}{\Lambda}]$ arrays for the truss elements are immediately obtained. For example, $[\overset{(3)}{\Lambda}]$ and $[\overset{(4)}{\Lambda}]$ are

$$\overset{(3)}{[\Lambda]} = \begin{bmatrix} 0 & 0 & 1 & 0 & 0 & 0 & 0 & 0 & 0 & 0 \\ 0 & 0 & 0 & 1 & 0 & 0 & 0 & 0 & 0 & 0 \\ 0 & 0 & 0 & 0 & 0 & 0 & 1 & 0 & 0 & 0 \\ 0 & 0 & 0 & 0 & 0 & 0 & 0 & 1 & 0 & 0 \end{bmatrix} \tag{2.66}$$

$$\overset{(4)}{[\Lambda]} = \begin{bmatrix} 0 & 0 & 1 & 0 & 0 & 0 & 0 & 0 & 0 & 0 \\ 0 & 0 & 0 & 1 & 0 & 0 & 0 & 0 & 0 & 0 \\ 0 & 0 & 0 & 0 & 0 & 0 & 0 & 0 & 1 & 0 \\ 0 & 0 & 0 & 0 & 0 & 0 & 0 & 0 & 0 & 1 \end{bmatrix} \tag{2.67}$$

In addition to Equation (2.64) which expresses the local displacements in terms of the global displacements of Equations (2.63), the incidence matrices may be used to express the global displacements in terms of local displacements. To do this, consider products of the form: $[\overset{(e)}{\Lambda}]^T \{\overset{e}{u}\}$. Specifically, consider as an example, $[\overset{(3)}{\Lambda}]^T \{\overset{(3)}{u}\}$:

$$
[\overset{(3)}{\Lambda}]^T \{\overset{3}{u}\} =
\begin{bmatrix}
0 & 0 & 0 & 0 \\
0 & 0 & 0 & 0 \\
1 & 0 & 0 & 0 \\
0 & 1 & 0 & 0 \\
0 & 0 & 0 & 0 \\
0 & 0 & 0 & 0 \\
0 & 0 & 1 & 0 \\
0 & 0 & 0 & 1 \\
0 & 0 & 0 & 0 \\
0 & 0 & 0 & 0
\end{bmatrix}
\begin{Bmatrix}
u_{1x}^{(3)} \\
u_{1y}^{(3)} \\
u_{2x}^{(3)} \\
u_{2y}^{(3)}
\end{Bmatrix}
=
\begin{Bmatrix}
0 \\
0 \\
u_{1x}^{(3)} \\
u_{1x}^{(3)} \\
0 \\
0 \\
u_{2x}^{(3)} \\
u_{2y}^{(3)} \\
0 \\
0
\end{Bmatrix}
\overset{D}{=} \{\overset{(3)}{U}\}
\tag{2.68}
$$

Since the resulting column array is not the complete global displacement array, it is labeled as $\{\overset{(3)}{U}\}$ and it can be considered as an expansion of the local displacement array into the "global system." Indeed, $\{\overset{(3)}{U}\}$ might be written

$$
\{\overset{(3)}{U}\} =
\begin{bmatrix}
0 \\
0 \\
u_{1x}^{(3)} \\
u_{1y}^{(3)} \\
0 \\
0 \\
u_{2x}^{(3)} \\
u_{2y}^{(3)} \\
0 \\
0
\end{bmatrix}
=
\begin{bmatrix}
0 \\
0 \\
U_{2x} \\
U_{2y} \\
0 \\
0 \\
U_{4x} \\
U_{4y} \\
0 \\
0
\end{bmatrix}
=
\begin{bmatrix}
0 \\
0 \\
U_3 \\
U_4 \\
0 \\
0 \\
U_7 \\
U_8 \\
0 \\
0
\end{bmatrix}
\tag{2.69}
$$

In general then, we can write

$$
[\overset{(e)}{\Lambda}]^T \{\overset{(e)}{u}\} = \{\overset{(e)}{U}\}
\tag{2.70}
$$

Similarly, in considering the local and global force arrays, we can write, in direct analogy to Equation (2.70),

$$[\overset{(e)}{\Lambda}]^T \{\overset{(e)}{f}\} = \{\overset{(e)}{F}\} \tag{2.71}$$

where $\{\overset{(e)}{F}\}$ is the truss element force array expanded into global dimensions. For example,

$$[\overset{(3)}{\Lambda}]^T \{\overset{(3)}{f}\} = \{\overset{(3)}{F}\} = \begin{bmatrix} 0 \\ 0 \\ f_{1x}^{(3)} \\ f_{1y}^{(3)} \\ 0 \\ 0 \\ f_{2x}^{(3)} \\ f_{2y}^{(3)} \\ 0 \\ 0 \end{bmatrix} \tag{2.72}$$

We can now relate the expanded force and displacement arrays to each other by recalling the fundamental relation of Equation (2.20):

$$\{\overset{(e)}{f}\} = [\overset{(e)}{k}]\{\overset{(e)}{u}\} \tag{2.73}$$

Hence, by Equations (2.64), (2.71), and (2.73)

$$\begin{aligned}
\{\overset{(e)}{F}\} &= [\overset{(e)}{\Lambda}]^T\{\overset{(e)}{f}\} = [\overset{(e)}{\Lambda}]^T[\overset{(e)}{k}]\{\overset{(e)}{u}\} \\
&= [\overset{(e)}{\Lambda}]^T[\overset{(e)}{k}][\overset{(e)}{\Lambda}]\{U\} \\
&\overset{D}{=} [\overset{(e)}{K}]\{U\}
\end{aligned} \tag{2.74}$$

where $\overset{(e)}{[K]}$ as defined by the last equality is the truss element stiffness matrix expanded into the global system. By adding Equations (2.74) for all elements, we have

$$\underset{(e)}{\Sigma} \overset{(e)}{\{F\}} = \underset{(e)}{\Sigma} (\overset{(e)}{[K]} \{U\}) = (\underset{(e)}{\Sigma} \overset{e}{[K]}) \{U\} \tag{2.75}$$

Let us now examine the sum of the forces on the left side of Equation (2.75) in detail. Specifically for the example truss of Figure 2.16,

$$\underset{(e)}{\overset{(e)}{\Sigma} \{F\}} =
\begin{bmatrix} f_{1x}^{(1)} \\ f_{1y}^{(1)} \\ 0 \\ 0 \\ 0 \\ 0 \\ f_{2x}^{(1)} \\ f_{2y}^{(1)} \\ 0 \\ 0 \end{bmatrix}
+
\begin{bmatrix} f_{1x}^{(2)} \\ f_{1y}^{(2)} \\ f_{2x}^{(2)} \\ f_{2y}^{(2)} \\ 0 \\ 0 \\ 0 \\ 0 \\ 0 \\ 0 \end{bmatrix}
+
\begin{bmatrix} 0 \\ 0 \\ f_{1x}^{(3)} \\ f_{1y}^{(3)} \\ 0 \\ 0 \\ f_{2x}^{(3)} \\ f_{2y}^{(3)} \\ 0 \\ 0 \end{bmatrix}
+
\begin{bmatrix} 0 \\ 0 \\ f_{1x}^{(4)} \\ f_{1y}^{(4)} \\ 0 \\ 0 \\ 0 \\ 0 \\ f_{2x}^{(4)} \\ f_{2y}^{(4)} \end{bmatrix}
+
\begin{bmatrix} 0 \\ 0 \\ f_{1x}^{(5)} \\ f_{1y}^{(5)} \\ f_{2x}^{(5)} \\ f_{2y}^{(5)} \\ 0 \\ 0 \\ 0 \\ 0 \end{bmatrix}
+
\begin{bmatrix} 0 \\ 0 \\ 0 \\ 0 \\ 0 \\ 0 \\ f_{1x}^{(6)} \\ f_{1y}^{(6)} \\ f_{2x}^{(6)} \\ f_{2y}^{(6)} \end{bmatrix}
+
\begin{bmatrix} 0 \\ 0 \\ 0 \\ 0 \\ f_{1x}^{(7)} \\ f_{1y}^{(7)} \\ 0 \\ 0 \\ f_{2x}^{(7)} \\ f_{2y}^{(7)} \end{bmatrix}
\tag{2.76}$$

or

$$\underset{(e)}{\Sigma} \overset{(e)}{\{F\}} =
\begin{bmatrix}
f_{1x}^{(1)} + f_{1x}^{(2)} \\
f_{1y}^{(1)} + f_{1y}^{(2)} \\
f_{2x}^{(2)} + f_{1x}^{(3)} + f_{1x}^{(4)} + f_{1x}^{(5)} \\
f_{2y}^{(2)} + f_{1y}^{(3)} + f_{1y}^{(4)} + f_{1y}^{(5)} \\
f_{2x}^{(5)} + f_{1x}^{(7)} \\
f_{2y}^{(5)} + f_{1y}^{(7)} \\
f_{2x}^{(1)} + f_{2x}^{(3)} + f_{1x}^{(6)} \\
f_{2y}^{(1)} + f_{2y}^{(3)} + f_{1y}^{(6)} \\
f_{2x}^{(4)} + f_{2x}^{(6)} + f_{2x}^{(7)} \\
f_{2y}^{(4)} + f_{2y}^{(6)} + f_{2y}^{(7)}
\end{bmatrix}
\tag{2.77}$$

From equilibrium considerations of free-body diagrams of the pins at the joints of the truss, we see that the elements of this column array are precisely the global force components on the joints. For example, a free-body diagram of joint 2 is shown in Figure 2.18. Hence, we can write

$$\sum_{(e)} \{\overset{(e)}{F}\} = \{F\} = \{P\} \tag{2.78}$$

where $\{P\}$ is the applied load array. (The loads are always to be applied to the joints.)

Finally, by comparing Equations (2.75) and (2.78),

$$\{F\} = (\sum_{(e)} \overset{(e)}{[K]}) \{U\} \tag{2.79}$$

Therefore, $\sum_{(e)} \overset{(e)}{[K]}$ may be identified as the global stiffness matrix $[K]$. That is,

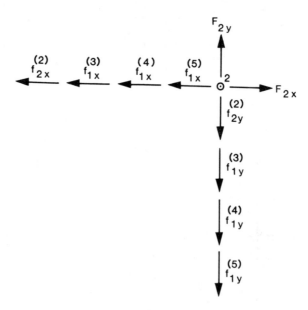

Figure 2.18   Free-Body Diagram of Joint 2

Table 2.4   Contributions to K_{IJ} from $k_{ij}^{(e)}$

i, j from $k_{ij}^{(e)}$	Truss Element e						
	1	2	3	4	5	6	7
1	1	1	3	3	3	7	5
2	2	2	4	4	4	8	6
3	7	3	7	9	5	9	9
4	8	4	8	10	6	10	10

I, J from K_{IJ}

$$[K] = \sum_{(e)} [\overset{(e)}{K}] \qquad\qquad (2.80)$$

The summation on the right side of Equation (2.80) may be obtained with the aid of Table 2.3. That is, by rewriting Table 2.3 as shown in Table 2.4, the contribution to $[K]$ from $k_{ij}^{(e)}$ is an entry in the I, J position obtained from the (e) column of the table. For example, $k_{13}^{(3)}$ is inserted into the 3, 7 position in $[K]$.

2.7   SUPPORT REACTIONS

In the foregoing analysis, if we substitute Equations (2.78) and (2.80) into (2.79) we obtain the following matrix governing equation of the example truss (Figure 2.19).

$$\{F\} = [K] \{U\} \qquad\qquad (2.81)$$

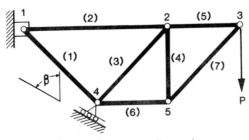

Figure 2.19   Truss Example

This matrix equation is equivalent to 10 scalar equations and hence, if we are given appropriate support conditions (that is, statically determinate), we can solve for a combination of 10 unknown displacements and reaction forces. In this section, we will examine various procedures for obtaining these solutions. We will illustrate these procedures by applying them with our example truss of Figure 2.19.

We begin by first observing that the load array may be divided into two parts as follows:

$$\{F\} = \{P\} + \{R\} \tag{2.82}$$

where $\{P\}$ consists of the externally applied forces and $\{R\}$ is the support reaction array. For the truss of Figure 2.19, the $\{P\}$ and $\{R\}$ arrays are

$$\{P\} = \begin{bmatrix} 0 \\ 0 \\ 0 \\ 0 \\ 0 \\ -P \\ 0 \\ 0 \\ 0 \\ 0 \end{bmatrix}, \quad \{R\} = \begin{bmatrix} R_1 \\ R_2 \\ 0 \\ 0 \\ 0 \\ 0 \\ R\cos\beta \\ R\sin\beta \\ 0 \\ 0 \end{bmatrix} \tag{2.83}$$

where R_1 and R_2 are the horizontal and vertical components of the pin reaction on joint 1 and where R is the resultant force exerted by the roller support at joint 4.

The pin support at joint 1 prevents both horizontal and vertical displacement and the roller support at joint 4 prevents displacement in a direction signified by a unit vector n perpendicular to the roller surface. These supports therefore constrain the displacement of the truss and they immediately produce the constraint equations

$$U_1 = U_2 = 0 \qquad (2.84)$$

and

$$U_7 \cos\beta + U_8 \sin\beta = 0 \qquad (2.85)$$

[Equation (2.85) is obtained by taking the scalar product of R with the vector displacement of joint 4.]

Equations (2.83), (2.84), and (2.85) could be regarded as being equivalent to a set of 13 scalar equations for the 10 inknown displacements U_i ($i = 1, \ldots, 10$) and the three unknown reaction forces R_1, R_2, and R. However, the constraint Equations (2.84) are trivial, providing values for U_1 and U_2 immediately. Hence, Equations (2.81) and (2.85) could be regarded as a system of 11 scalar equations for the 11 unknowns U_i ($i = 3, \ldots, 10$), R_1, R_2 and R. There are several approaches that can be used to formally incorporate Equation (2.85) (and hence, constraint equations like it) into the matrix system of Equation (2.81). An obvious approach is simply to solve Equation (2.85) for, say, U_7 in terms of U_8 and then to replace U_7 in Equations (2.81) giving a system of 10 equations and 10 unknowns. Although this approach will be satisfactory in our relatively simple example (providing $\beta \neq 90°$), it is not easily generalized for more complex configurations, nor is it easily automated into a computer algorithm.

A second approach is to remodel the roller support by replacing it with a very stiff element perpendicular to the roller support surface as shown in Figure 2.20. Although this approach has the disadvantage of introducing another element [and hence, increasing the array dimensions of Equations (2.81)], it has the advantage of having all the constraint equations of the simple, trivial form of Equations (2.84). A more subtle disadvantage, however, is that with the stiffness of element 8 much greater than the stiffness of the other elements, the numerical solution of Equations (2.81) could become more difficult due to convergence problems.

A third approach (an approach which is used in many truss analysis computer programs) is to introduce a coordinate trans-

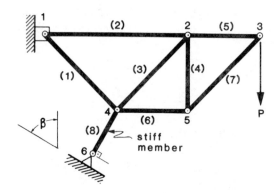

Figure 2.20   Replacement of Roller Support
at Joint 4 with a Stiff Element

formation at joint 4 so that the displacement axes are, respectively, parallel and perpendicular to the roller support. Specifically, let us introduce new axes, $\hat{X}$ and $\hat{Y}$, as shown in Figure 2.21. Then the global joint displacements U_7 and U_8 may be written as

$$U_7 = S_\beta \, \hat{U}_7 + C_\beta \, \hat{U}_8$$
$$U_8 = -C_\beta \, \hat{U}_7 + S_\beta \, \hat{U}_8$$

(2.86)

where U_7 and U_8 are the global joint displacements along X and Y, respectively, and, as before, S_β and C_β are abbreviations for $\sin\beta$ and $\cos\beta$. These equations may be written in the matrix form

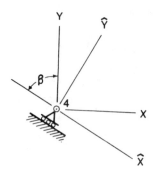

Figure 2.21   Coordinate Axes at Joint 4

$$
\begin{bmatrix} U_7 \\ U_8 \end{bmatrix} = \begin{bmatrix} S_\beta & C_\beta \\ -C_\beta & S_\beta \end{bmatrix} \begin{bmatrix} \hat{U}_7 \\ \hat{U}_8 \end{bmatrix} \tag{2.87}
$$

Thus, let us introduce a new global displacement array $\{\hat{U}\}$, which is the same as $\{U\}$, except that U_7 and U_8 are replaced by $\hat{U}_7$ and $\hat{U}_8$. Then $\{U\}$ and $\{\hat{U}\}$ are related as follows:

$$
\begin{bmatrix} U_1 \\ U_2 \\ U_3 \\ U_4 \\ U_5 \\ U_6 \\ U_7 \\ U_8 \\ U_9 \\ U_{10} \end{bmatrix} = \begin{bmatrix} 1 & & & & & & & & & \\ & 1 & & & & & & & & \\ & & 1 & & & 0 & & & & \\ & & & 1 & & & & & & \\ & & & & 1 & & & & & \\ & & & & & 1 & & & & \\ & & & & & & S_\beta & C_\beta & & \\ & & 0 & & & & -C_\beta & S_\beta & & \\ & & & & & & & & 1 & \\ & & & & & & & & & 1 \end{bmatrix} \begin{bmatrix} U_1 \\ U_2 \\ U_3 \\ U_4 \\ U_5 \\ U_6 \\ \hat{U}_7 \\ \hat{U}_8 \\ U_9 \\ U_{10} \end{bmatrix} \tag{2.88}
$$

That is,

$$
\{U\} = [T]\{\hat{U}\} \tag{2.89}
$$

where $[T]$ is defined by Equation (2.88). Similarly, a new support reaction array $\{\hat{R}\}$ may be introduced as follows:

$$
\{\hat{R}\} = \begin{bmatrix} R_1 \\ R_2 \\ 0 \\ 0 \\ 0 \\ 0 \\ \hat{R}_7 \\ \hat{R}_8 \\ 0 \\ 0 \end{bmatrix} \tag{2.90}
$$

where $\hat{R}_7$ and $\hat{R}_8$ are the components of the roller support reactions in the $\hat{X}$ and $\hat{Y}$ directions. However, from the physical nature of a roller support we have

$$\hat{R}_7 = 0 \tag{2.91}$$

and

$$\hat{R}_8 = R \tag{2.92}$$

[See Equations (2.83).] $\{R\}$ and $\{\hat{R}\}$ are related in exactly the same manner as $\{U\}$ and $\{\hat{U}\}$. That is,

$$\{R\} = [T] \{\hat{R}\} \tag{2.93}$$

By substituting Equations (2.82), (2.89), and (2.93) into Equation (2.81) we obtain

$$[K] [T] \{\hat{U}\} = \{P\} + [T]\{\hat{R}\} \tag{2.94}$$

Since $[T]$ is orthogonal, that is, the transpose is the inverse, Equation (2.94) may be rewritten

$$[\hat{K}]\{\hat{U}\} = \{\hat{P}\} + \{\hat{R}\} \tag{2.95}$$

where

$$[\hat{K}] = [T]^T [K] [T] \tag{2.96}$$

and

$$\{\hat{P}\} = [T]^T \{P\} \tag{2.97}$$

Now, in our example, $\{\hat{P}\}$ is the same as $\{P\}$ and the constraint equations are of the simple form

$$U_1 = U_2 = \hat{U}_8 = 0 \tag{2.98}$$

and the unknown support reaction force components are R_1, R_2, and $\hat{R}_8$. Equations (2.95) are thus equivalent to a system of 10 scalar equations for the 7 unknown displacement components U_3, U_4, U_5, U_6, $\hat{U}_7$, U_9 and U_{10} and the three unknown force components R_1, R_2 and $\hat{R}_8$.

This approach (which is perhaps the most popular of all the approaches with constraint equations and reaction forces) has the advantage of being computer-oriented and of using the "natural" coordinates or components of the physical system. A disadvantage with this approach is that it requires a number of matrix multiplications, such as in Equations (2.94), (2.96) and (2.97). This could be a significant factor with very large trusses. (However, see Problem 2.11.)

There may be occasions in the solution of Equations (2.81) with constraints such as Equations (2.84) and (2.85) when we are primarily interested in knowing the displacements as opposed to the reaction forces. In this case, it is desirable to reduce the system to a set of 7 scalar equations for the 7 unknown displacements. The following approach accomplishes this objective.

Consider writing the constraint Equations (2.84) and (2.85) in the matrix form

$$
\begin{bmatrix}
1 & 0 & 0 & 0 & 0 & 0 & 0 & 0 & 0 & 0 \\
0 & 1 & 0 & 0 & 0 & 0 & 0 & 0 & 0 & 0 \\
0 & 0 & 0 & 0 & 0 & 0 & \cos\beta & \sin\beta & 0 & 0
\end{bmatrix}
\begin{bmatrix}
U_1 \\ U_2 \\ U_3 \\ U_4 \\ U_5 \\ U_6 \\ U_7 \\ U_8 \\ U_9 \\ U_{10}
\end{bmatrix} = 0 \quad (2.99)
$$

or simply

$$[A]\{U\} = 0 \tag{2.100}$$

where the matrix $[A]$ is defined by Equation (2.99). The support reaction array may then be written

$$\{R\} = \begin{bmatrix} 1 & 0 & 0 \\ 0 & 1 & 0 \\ 0 & 0 & 0 \\ 0 & 0 & 0 \\ 0 & 0 & 0 \\ 0 & 0 & 0 \\ 0 & 0 & \cos\beta \\ 0 & 0 & \sin\beta \\ 0 & 0 & 0 \\ 0 & 0 & 0 \end{bmatrix} \begin{Bmatrix} R_1 \\ R_2 \\ R \end{Bmatrix} \qquad (2.101)$$

or simply

$$\{R\} = [A]^T \{S\} \qquad (2.102)$$

where $\{S\}$ is defined as the array of 3 unknown forces of Equation (2.101). By using Equations (2.82) and (2.102), the governing equations (2.81) may be written

$$[K]\{U\} = \{P\} + [A]^T\{S\} \qquad (2.103)$$

Next, introduce 7 arrays $\{\eta\}_i$ $(i = 1, \ldots, 7)$ which are "orthogonal" to $[A]^T \{S\}$, or $\{R\}$, as follows

$$\{\eta\}_1 = \begin{bmatrix} 0 \\ 0 \\ 1 \\ 0 \\ 0 \\ 0 \\ 0 \\ 0 \\ 0 \\ 0 \end{bmatrix}, \{\eta\}_2 = \begin{bmatrix} 0 \\ 0 \\ 0 \\ 1 \\ 0 \\ 0 \\ 0 \\ 0 \\ 0 \\ 0 \end{bmatrix}, \{\eta\}_3 = \begin{bmatrix} 0 \\ 0 \\ 0 \\ 0 \\ 1 \\ 0 \\ 0 \\ 0 \\ 0 \\ 0 \end{bmatrix}, \{\eta\}_4 = \begin{bmatrix} 0 \\ 0 \\ 0 \\ 0 \\ 0 \\ 1 \\ 0 \\ 0 \\ 0 \\ 0 \end{bmatrix}, \{\eta\}_5 = \begin{bmatrix} 0 \\ 0 \\ 0 \\ 0 \\ 0 \\ 0 \\ -\sin\beta \\ \cos\beta \\ 0 \\ 0 \end{bmatrix}, \{\eta\}_6 = \begin{bmatrix} 0 \\ 0 \\ 0 \\ 0 \\ 0 \\ 0 \\ 0 \\ 0 \\ 1 \\ 0 \end{bmatrix}, \{\eta\}_7 = \begin{bmatrix} 0 \\ 0 \\ 0 \\ 0 \\ 0 \\ 0 \\ 0 \\ 0 \\ 0 \\ 1 \end{bmatrix}$$

$$(2.104)$$

Then by premultiplying Equation (2.99) by $\{\eta_i\}^T$ ($i = 1, \ldots, 7$), we have

$$\{\eta\}_i^T [K] \{U\} = \{\eta\}_i^T \{P\} + \{\eta\}_i^T [A]^T \{S\} \quad i = 1, \ldots, 7 \quad (2.105)$$

Equations (2.105) form the desired set of 7 scalar equations for the 7 unknown displacements.

This approach has the advantage of reducing the dimension of the system of equations, but it has the disadvantage of requiring knowledge of the 7 arrays which are orthogonal to $[A]^T \{S\}$. In a more general problem, these arrays may be difficult to obtain.

We conclude this section by outlining a fifth approach for working with constraint equations and support reactions. This approach is very general and it is therefore convenient with situations where the constraints are more complex than those of our example problem. We encounter such situations in finite-element analysis of other kinds of structures and systems.

Suppose, in general, we have the following matrix system, such as Equations (2.81) which is equivalent to n scalar equations

$$[K]\{U\} = \{P\} + [A]^T \{S\} \tag{2.106}$$

Suppose further that the constraint equations, such as Equations (2.84) and (2.85), are equivalent to m ($m < n$) scalar equations and that they are written in the matrix form

$$[A]\{U\} = 0 \tag{2.107}$$

In these equations, let the variables be ordered so that we can solve Equations (2.107) for, say, the last m of the U_i in terms of the first $n - m$ of the U_i. Hence, let $[A]$ and $\{U\}$ of Equation (2.107) be partitioned as follows:

$$[A_1 \mid A_2] \left\{ \frac{\tilde{U}_1}{\tilde{U}_2} \right\} = 0 \tag{2.108}$$

where $\{\tilde{U}_1\}$ contains the first $n - m$ of the U_i and $\{\tilde{U}_2\}$ contains the remaining m U_i.

Now, assuming $[A_2]^{-1}$ exists, we can solve Equation (2.108) for $\{\tilde{U}_2\}$ as

$$\{\tilde{U}_2\} = -[A_2]^{-1}[A_1]\{\tilde{U}_1\} \qquad (2.109)$$

As with Equation (2.107), Equation (2.106) may be similarly partitioned as

$$\left[\begin{array}{c|c}\tilde{K}_{11} & \tilde{K}_{12} \\ \hline \tilde{K}_{21} & \tilde{K}_{22}\end{array}\right]\left\{\begin{array}{c}\tilde{U}_1 \\ \hline \tilde{U}_2\end{array}\right\} = \left\{\begin{array}{c}\tilde{P}_1 \\ \hline \tilde{P}_2\end{array}\right\} + \left[\begin{array}{c}A_1^T \\ \hline A_2^T\end{array}\right]\{S\} \qquad (2.110)$$

or

$$\overset{n-m}{\underset{n-m}{}}[\tilde{K}_{11}]\overset{1}{\{\tilde{U}_1\}} + \overset{m}{}[\tilde{K}_{12}]\overset{1}{\{\tilde{U}_2\}} = \overset{1}{\{\tilde{P}_1\}} + \overset{m}{[A_1^T]}\{S\} \qquad (2.111)$$

and

$$\overset{n-m}{\underset{m}{}}[\tilde{K}_{21}]\overset{1}{\{\tilde{U}_1\}} + \overset{m}{}[\tilde{K}_{22}]\overset{1}{\{\tilde{U}_2\}} = \overset{1}{\{\tilde{P}_2\}} + \overset{m}{[A_2^T]}\{S\} \qquad (2.112)$$

where the matrix sizes are indicated. [Equations (2.111) and (2.112) are obtained by simple matrix multiplication of the partitioned matrices of (2.110).] By substituting for $\{\tilde{U}_2\}$ from Equation (2.109), these equations become

$$[\tilde{K}_{11}]\{\tilde{U}_1\} - [\tilde{K}_{12}][A_2]^{-1}[A_1]\{\tilde{U}_1\} = \{\tilde{P}_1\} + [A_1^T]\{S\} \qquad (2.113)$$

and

$$[\tilde{K}_{21}]\{\tilde{U}_1\} - [\tilde{K}_{22}][A_2]^{-1}[A_1]\{\tilde{U}_1\} = \{\tilde{P}_2\} + [A_2^T]\{S\} \qquad (2.114)$$

By solving Equation (2.114) for $\{S\}$, we obtain

$$\{S\} = -[A_2^T]^{-1}\{\tilde{P}_2\}$$
$$+ [A_2^T]^{-1}([\tilde{K}_{21}] - [\tilde{K}_{22}][A_2]^{-1}[A_1])\{\tilde{U}_1\} \qquad (2.115)$$

Finally, substituting for $\{S\}$ in Equation (2.113) leads to

$$\langle[\tilde{K}_{11}] - [\tilde{K}_{12}] [A_2]^{-1} [A_1] - [A_1^T] [A_2^T]^{-1} ([\tilde{K}_{21}]$$

$$(2.116)$$

$$- [\tilde{K}_{22}] [A_2]^{-1} [A_1])\rangle \{\tilde{U}_1\} = \{\tilde{P}_1\} - [A_1^T] [A_2^T]^{-1} \{P_2\}$$

This equation may be written in the more compact form

$$[K] \{\tilde{U}_1\} = \{\rho\} \qquad (2.117)$$

where $[K]$ and $\{\rho\}$ are defined by inspection from Equation (2.116).

In Equation (2.117), $[K]$ and $\{\rho\}$ play the role of modified stiffness and load arrays.

The solution procedure is simply to solve Equations (2.117) for the $n - m$ unknown U_i in $\{\tilde{U}_1\}$. Then, the remaining U_i, of $\{\tilde{U}_2\}$, together with the support reaction forces of $\{S\}$ may be obtained from Equations (2.111) and (2.115).

2.8 COMPUTER SOLUTIONS

Although the foregoing analysis has been developed through illustration with a relatively simple truss, it may be applied following the outlined procedures, to any truss. Moreover, the analysis is developed so that algorithms are easily written for an automated development of the governing equations. Indeed, the governing equations

$$[K] \{U\} = \{L\} + \{R\} \qquad (2.118)$$

are ideally suited for development and solution on a high-speed digital computer. Specifically, the following procedure might be used:

1. For a given truss, label the elements and joints
2. Write an algorithm to develop and store the element-joint identification tables such as Tables 2.1, 2.2, 2.3, and 2.4 of the example truss

3. Write an algorithm to compute the incidence matrices $[\Lambda]$, from a table such as Table 2.3

4. Develop a subroutine to calculate the element stiffness matrices

5. Write an algorithm to calculate the global stiffness matrix $[K]$ from the element stiffness matrices following the procedure of Table 2.4

6. Finally, solve the governing equations for the unknown displacement and reaction forces using one of the approaches of the previous section for the constraint equations.

Many computer programs for analyzing trusses have been written following this basic procedure. Indeed, a number are commercially available. See also Problem 2.12.

2.9   THREE-DIMENSIONAL TRUSSES

We conclude this chapter with a few remarks about three-dimensional trusses. The analysis of the foregoing sections was, of course, developed for two-dimensional or plane trusses. However, three-dimensional trusses may be analyzed following exactly the same procedures. Indeed, the only modification required in the foregoing theory is to extend the arrays to include the third dimension's parameters, that is, the force, displacement, and stiffness parameters. Specifically, the element force and displacement arrays become

$$\{f\} = \begin{bmatrix} f_{1x} \\ f_{1y} \\ f_{1z} \\ f_{2x} \\ f_{2y} \\ f_{2z} \end{bmatrix} \qquad \{u\} = \begin{bmatrix} u_{1x} \\ u_{1y} \\ u_{1z} \\ u_{2x} \\ u_{2y} \\ u_{2z} \end{bmatrix} \qquad (2.119)$$

The element stiffness array [analogous to Equation (2.21)], becomes

$$[k] = k \begin{bmatrix} c_x^2 & c_x c_y & c_x c_z & -c_x^2 & -c_x c_y & -c_x c_z \\ c_x c_y & c_y^2 & c_y c_z & -c_x c_y & -c_y^2 & -c_y c_z \\ c_x c_z & c_y c_z & c_z^2 & -c_x c_z & -c_y c_z & -c_z^2 \\ -c_x^2 & -c_x c_y & -c_x c_z & c_x^2 & c_x c_y & c_x c_z \\ -c_x c_y & -c_y^2 & -c_y c_z & c_x c_y & c_y^2 & c_y c_z \\ -c_x c_z & -c_y c_z & -c_z^2 & c_x c_z & c_y c_z & c_z^2 \end{bmatrix} \qquad (2.120)$$

where c_x, c_y, and c_z are direction cosines of the element relative to the X, Y, and Z directions. The assembly procedure, the development of the global stiffness matrix, the analysis with the constraint equations, and the computer formulation directly follow the procedures outlined in the foregoing sections.

PROBLEMS

Section 2.2

2.1 The element k_{ij} of the stiffness matrix is sometimes referred to as "the force at end i due to a unit displacement at end j." Check that this description is valid for the elements of the stiffness matrix of Equation (2.10). *Hint*: Use Equations (2.6) and (2.7) together with Figures 2.3 and 2.4.

Section 2.4

2.2 Verify Equations (2.33) and (2.36)

2.3 Expand the matrix product of Equation (2.43) and verify that the result is the stiffness matrix of Equation (2.25).

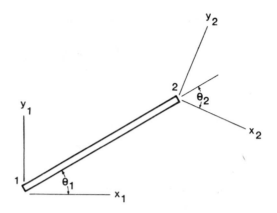

Figure P2.4 Truss Element with Different
Global Directions at its Ends

2.4 Using the procedures of this section, develop a stiffness matrix for an element that is inclined to different global axes at its ends as shown in Figure P2.4.

Result:

$$
[k] = k
\begin{bmatrix}
c_1^2 & s_1 c_1 & -c_1 c_2 & -c_1 s_2 \\
s_1 c_1 & s_1^2 & -s_1 c_2 & -s_1 s_2 \\
-c_1 c_2 & -s_1 c_2 & c_2^2 & s_2 c_2 \\
-s_2 c_1 & -s_1 s_2 & s_2 c_2 & s_2^2
\end{bmatrix}
$$

where s_1, c_1, s_2, and c_2 represent $\sin\theta$, $\cos\theta$, $\sin\theta_2$, and $\cos\theta_2$, respectively.

Section 2.5

2.5 Perform the substitution of Equations (2.52) through (2.55) into Equation (2.56) and thus verify Equation (2.57).

2.6 For the global stiffness matrix $[K]$ of Equation (2.57) verify that: (a) it is symmetric; (b) the sum of its elements in any row or column is zero; and (c) it is singular.

2.7 Find the unknown forces and displacements for the simple truss of Figure 2.11 by using the methods of elementary statics and strength of materials and thus verify the results of Equation (2.61). Suggested procedure: Use the method of joints to find the forces in each member. Then from Hooke's law, find the elongation of each number. Finally, compute the displacement of each joint by noting that the elongation of a member is simply the projection (or component) of relative displacement of its ends, along the member itself.

Section 2.6

2.8 Using Table 2.3, find $[\overset{(2)}{\Lambda}]$, $[\overset{(5)}{\Lambda}]$, $[\overset{(6)}{\Lambda}]$, and $[\overset{(7)}{\Lambda}]$ for the example truss of Figures 2.16 and 2.17.

2.9 Verify, by examination of $[\overset{(e)}{\Lambda}]$ and $[\overset{(e)}{\Lambda}]^T$, that the procedure outlined following Equation (2.80) for calculating the global stiffness matrix $[K]$ from the element stiffness matrices $[\overset{(e)}{k}]$, is correct.

Section 2.7

2.10 Show that the only terms of the global stiffness matrix $[K]$ affected by the transformation of Equation (2.93) are the terms associated with joint 4 of the example truss.

2.11 See Problem 2.10. Show that the global stiffness matrix $[\hat{K}]$ of Equation (2.96) could have been obtained by using the $\hat{X}$ and $\hat{Y}$ axes with the ends of elements 1, 3, and 6 which are connected at joint 4.

2.12 As a project, develop a computer program following the procedures outlined in Section 2.8 for the analysis of plane and/or space trusses.

3

Beams and Frames

3.1 INTRODUCTION

In this chapter we will apply the finite element method in the analysis of beams and frames.* It will be seen that this is simply a natural extension of the procedures developed with trusses in Chapter 2. Indeed, we will employ the same general methods. Moreover, we will use the same tutorial approach by developing the analysis and procedures through an example problem. Specifically, we will make an analysis of the frame shown in Figure 3.1. That is, we will seek to determine the displacements and beam rotations at the joints and the corresponding forces and bending moments. However, as with the truss example of Chapter 2, our main objective is to develop a general analysis with procedures which will be applicable with a broad range of beam-frame systems.

The basis of the finite element approach with beam-frame systems, as with trusses, is the division or separation of the structure

---

*By a "frame," we mean a structure consisting of welded and pinned beams.

71

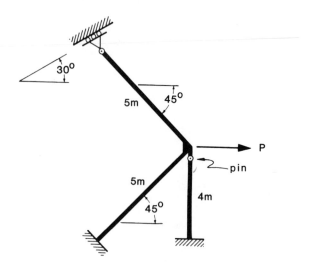

Figure 3.1   Example Frame

into segments or elements. Following this segmentation, a force-displacement analysis is performed on each individual element. The global effect on the complete structure is then obtained by super-position following assembly procedures similar to those in Chapter 3. Therefore, we begin our analysis by considering beam elements.

3.2 BEAM ELEMENTS

At this point, it might be helpful for those readers who are unfamiliar with or who have forgotten the basic ideas of elementary beam theory, to review them in one of the many strength-of-materials texts. In our analysis, we will impose on our beam-frame structures and thus, on our beam elements, all the assumptions (for example: small displacements, thin beams, and plane sections normal to the neutral axis before bending remain plane and normal to the neutral axis after bending, etc.) needed for elementary beam theory to be valid. Beyond this we will adopt the following sign convention for our beam elements: Forces and displacements in the positive horizontal (to the right) and positive vertical (upward) directions are

Figure 3.2 Positive Force and Moment Directions

positive at each end of the element. Also, moments and rotations in the counterclockwise direction are positive at each end. These are illustrated in Figures 3.2 and 3.3. (It should be noted that some writers on beam theory have adopted different sign conventions.)

To begin the details of our analysis, we recall the following results from the elementary theory of cantilever beams: For a beam with a single concentrated vertical load with magnitude f_{2y} on its right end (see Figure 3.4), the vertical displacement u_{2y}, and end rotation θ_2 are

$$u_{2y} = f_{2y}\ell^3/3EI \tag{3.1}$$

and

$$\theta_2 = f_{2y}\ell^2/2EI \tag{3.2}$$

where ℓ is the beam length, E is Young's modulus of elasticity, and I is the second moment of area of the beam cross section about its neutral axis.

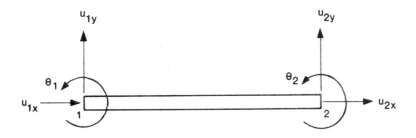

Figure 3.3 Positive Displacement and Rotation Directions

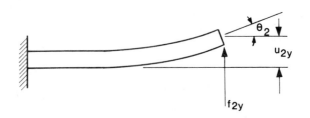

Figure 3.4 Cantilever Beams with Concentrated End Load

Similarly, for a beam with concentrated moment m_2 at its end (see Figure 3.5), the rotation θ_2 at the end of the beam and vertical displacement are

$$\theta_2 = m_2 \ell/EI \tag{3.3}$$

and

$$u_{2y} = m_2 \ell^2/2EI \tag{3.4}$$

Finally, as we have already noted in Chapter 2, if a cantilever beam is subjected to a single horizontal load with magnitude f_{2x} (see Figure 3.6), the horizontal displacement u_{2x} of the end is

$$u_{2x} = f_{2x} \ell/AE \tag{3.5}$$

where A is the beam cross-sectional area.

These results will be helpful in constructing the element stiffness matrix.

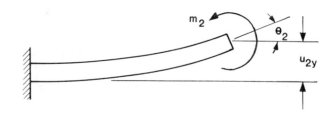

Figure 3.5 Cantilever Beam with Concentrated End Moment

Figure 3.6 Cantilever Beam with Horizontal End Load

3.3 ELEMENT STIFFNESS MATRIX

From Figure 3.2 we see that at each end of the beam element there are two forces and a moment. Also, from Figure 3.3 we see that at each end there are correspondingly, two displacements and a rotation. Hence, there are a total of six loading components and six deformation components. These are to be related to each other through the 6 × 6 element stiffness matrix $[k]$ as follows:

$$\{f\} = [k]\,\{u\} \tag{3.6}$$

where the element force array $\{f\}$ is

$$\{f\} = \begin{bmatrix} f_{1x} \\ f_{1y} \\ m_1 \\ f_{2x} \\ f_{2y} \\ m_2 \end{bmatrix} \tag{3.7}$$

and the element displacement array $\{u\}$ is

$$\{u\} = \begin{bmatrix} u_{1x} \\ u_{1y} \\ \theta_1 \\ u_{2x} \\ u_{2y} \\ \theta_2 \end{bmatrix} \tag{3.8}$$

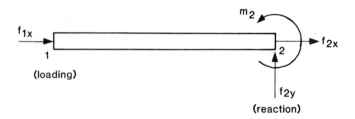

Figure 3.7 Beam Element Fixed at the Right End
with Axial Load on the Left End

The element stiffness matrix $[k]$ may be obtained by using the cantilever beam relations of the foregoing section together with the principle of superposition. Specifically, consider the beam element as shown in Figure 3.7 where the deformation at end 2 is held at zero and there is a horizontal force at end 1. Equilibrium considerations together with Equation (3.5) lead to the following expressions for the reaction forces and moment at end 2 and the displacements and rotation at end 1.

$$f_{2x} = -f_{1x}$$

$$f_{2y} = 0 \tag{3.9}$$

$$m_2 = 0$$

and

$$u_{1x} = f_{1x}\ell/AE$$

$$u_{1y} = 0 \tag{3.10}$$

$$\theta_1 = 0$$

Next, consider the beam element shown in Figure 3.8 with zero deformation at end 2 and a vertical force at end 1. Equilibrium considerations together with Equation (3.1) lead to the following expressions for the reactions at end 2 and the displacements at end 1:

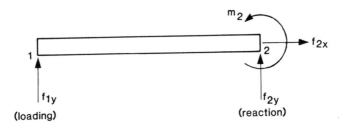

Figure 3.8 Beam Element Fixed at the Right End
with Vertical Load on the Left End

$$f_{2y} = -f_{1y}$$

$$m_2 = f_{1y}\ell \tag{3.11}$$

$$f_{2x} = 0$$

and

$$u_{1x} = 0$$

$$u_{1y} = f_{1y}\ell^3/3EI \tag{3.12}$$

$$\theta_1 = -f_{1y}\ell^2/2EI$$

Finally, consider the beam element shown in Figure 3.9 with zero deformation at end 2 and a concentrated moment at end 1.

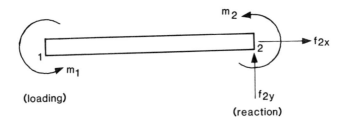

Figure 3.9 Beam Element Fixed at the Right End
with a Moment at the Left End

Equilibrium considerations together with Equation (3.3) lead to the following expressions for the reactions at end 2 and the displacements at end 1:

$$f_{2x} = 0$$
$$f_{2y} = 0 \tag{3.13}$$
$$m_2 = -m_1$$

and

$$u_{1x} = 0$$
$$u_{1y} = -m_1 \ell^2/2EI \tag{3.14}$$
$$\theta_1 = m_1 \ell/EI$$

If we superpose these three loading configurations, we have a beam element with zero deformation at end 2 but with a combined horizontal force, vertical force, and moment at end 1 as shown in Figure 3.10. Combining Equations (3.9), (3.11), and (3.13) leads to the following equilibrium relations for the reactions:

$$f_{2x} = -f_{1x}$$
$$f_{2y} = -f_{1y} \tag{3.15}$$
$$m_2 = f_{1y}\ell - m_1$$

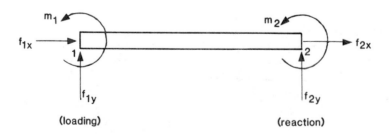

(loading)   (reaction)

Figure 3.10 Beam Element Fixed at the Right End
with Combined Loading on the Left End

In matrix form, these equations are

$$
\begin{bmatrix} f_{2x} \\ f_{2y} \\ m_2 \end{bmatrix} = \begin{bmatrix} -1 & 0 & 0 \\ 0 & -1 & 0 \\ 0 & \ell & -1 \end{bmatrix} \begin{bmatrix} f_{1x} \\ f_{1y} \\ m_1 \end{bmatrix} \tag{3.16}
$$

Similarly, combining Equations (3.10), (3.12), and (3.14) lead to the following deformation-loading relations:

$$
\begin{aligned}
u_{1x} &= f_{1x}\ell/AE \\
u_{1y} &= f_{1y}\ell^3/3EI - m_1\ell^2/2EI \\
\theta_1 &= -f_{1y}\ell^2/2EI + m_1\ell/EI
\end{aligned} \tag{3.17}
$$

In matrix form, these equations are

$$
\begin{bmatrix} u_{1x} \\ u_{1y} \\ \theta_1 \end{bmatrix} = \begin{bmatrix} \ell/AE & 0 & 0 \\ 0 & \ell^3/3EI & -\ell^2/2EI \\ 0 & -\ell^2/2EI & \ell/EI \end{bmatrix} \begin{bmatrix} f_{1x} \\ f_{1y} \\ m_1 \end{bmatrix} \tag{3.18}
$$

Recall, however, that our objective here is to express the loading or force array in terms of the deformation or displacement array and thus determine the stiffness matrix as in Equation (3.6). Therefore, by solving Equation (3.18) for f_{1x}, f_{1y}, and m_1, we have

$$
\begin{bmatrix} f_{1x} \\ f_{1y} \\ m_1 \end{bmatrix} = \begin{bmatrix} AE/\ell & 0 & 0 \\ 0 & 12EI/\ell^3 & 6EI/\ell^2 \\ 0 & 6EI/\ell^2 & 4EI/\ell \end{bmatrix} \begin{bmatrix} u_{1x} \\ u_{1y} \\ \theta_1 \end{bmatrix} \tag{3.19}
$$

Also, by substituting this expression into Equation (3.16), we have

$$
\begin{bmatrix} f_{2x} \\ f_{2y} \\ m_2 \end{bmatrix} = \begin{bmatrix} -AE/\ell & 0 & 0 \\ 0 & -12EI/\ell^3 & -6EI/\ell^2 \\ 0 & 6EI/\ell^2 & 2EI/\ell \end{bmatrix} \begin{bmatrix} u_{1x} \\ u_{1y} \\ \theta_1 \end{bmatrix} \tag{3.20}
$$

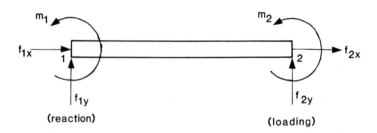

Figure 3.11   Beam Element Fixed at the Left End
with Combined Loading on the Right End

In a directly analogous manner to the above analysis, if we impose zero deformation on end 1 of the beam element and a combined horizontal force, vertical force, and moment on end 2 as shown in Figure 3.11, equilibrium considerations lead to the following expressions for the reactions at end 1 in terms of the loading at end 2:

$$f_{1x} = -f_{2x}$$

$$f_{1y} = -f_{2y} \tag{3.21}$$

$$m_1 = -f_{2y}\ell - m_2$$

In matrix form, these equations are

$$
\begin{bmatrix} f_{1x} \\ f_{1y} \\ m_1 \end{bmatrix} =
\begin{bmatrix} -1 & 0 & 0 \\ 0 & -1 & 0 \\ 0 & -\ell & -1 \end{bmatrix}
\begin{bmatrix} f_{2x} \\ f_{2y} \\ m_2 \end{bmatrix} \tag{3.22}
$$

Also, adapting Equations (3.1), (3.3), and (3.5) to the beam element of Figure 3.11 leads to the following expressions for the displacements at end 2 in terms of the loading:

$$u_{2x} = f_{2x}\ell/AE$$

$$u_{2y} = f_{2y}\ell^3/3EI + m_2\ell^2/2EI \tag{3.23}$$

$$\theta_2 = f_{2y}\ell^2/2EI + m_2\ell/EI$$

In matrix form, these equations are

$$
\begin{bmatrix} u_{2x} \\ u_{2y} \\ \theta_2 \end{bmatrix} = \begin{bmatrix} \ell/AE & 0 & 0 \\ 0 & \ell^3/3EI & \ell^2/2EI \\ 0 & \ell^2/2EI & \ell/EI \end{bmatrix} \begin{bmatrix} f_{2x} \\ f_{2y} \\ m_2 \end{bmatrix} \tag{3.24}
$$

Solving for f_{2x}, f_{2y}, and m_2, we have

$$
\begin{bmatrix} f_{2x} \\ f_{2y} \\ m_2 \end{bmatrix} = \begin{bmatrix} AE/\ell & 0 & 0 \\ 0 & 12EI/\ell^3 & -6EI/\ell^2 \\ 0 & -6EI/\ell^2 & 4EI/\ell \end{bmatrix} \begin{bmatrix} u_{2x} \\ u_{2y} \\ \theta_2 \end{bmatrix} \tag{3.25}
$$

Substituting this expression into Equation (3.22) we have

$$
\begin{bmatrix} f_{1x} \\ f_{1y} \\ m_1 \end{bmatrix} = \begin{bmatrix} -AE/\ell & 0 & 0 \\ 0 & -12EI/\ell^3 & 6EI/\ell^2 \\ 0 & -6EI/\ell^2 & 2EI/\ell \end{bmatrix} \begin{bmatrix} u_{2x} \\ u_{2y} \\ \theta_2 \end{bmatrix} \tag{3.26}
$$

Equations (3.19), (3.20), (3.25), and (3.26) contain the information needed to determine the desired element stiffness matrix $[k]$ of Equation (3.6). Indeed, if the element force and displacement arrays of Equations (3.7) and (3.8) are partitioned as follows:

$$
\{f\} = \begin{bmatrix} f_{1x} \\ f_{1y} \\ m_1 \\ -- \\ f_{2x} \\ f_{2y} \\ m_2 \end{bmatrix} = \begin{bmatrix} f_1 \\ -- \\ f_2 \end{bmatrix} \tag{3.27}
$$

and

$$\{u\} = \begin{bmatrix} u_{1x} \\ u_{1y} \\ \theta_1 \\ -- \\ u_{2x} \\ u_{2y} \\ \theta_2 \end{bmatrix} = \begin{bmatrix} u_1 \\ -- \\ u_2 \end{bmatrix} \tag{3.28}$$

Then Equation (3.6) may be written

$$\begin{bmatrix} f_1 \\ -- \\ f_2 \end{bmatrix} = \begin{bmatrix} \kappa_{11} & \vline & \kappa_{12} \\ --+-- \\ \kappa_{21} & \vline & \kappa_{22} \end{bmatrix} \begin{bmatrix} u_1 \\ -- \\ u_2 \end{bmatrix} \tag{3.29}$$

where the element striffness matrix $[k]$ is partitioned into four 3×3 matrices κ_{ij} $(i, j = 1, 2)$. Inspection of Equations (3.19), (3.20), (3.25), and (3.26), however, shows that these matrices are respectively the square matrices of these equations. That is,

$$[\kappa_{11}] = \begin{bmatrix} AE/\ell & 0 & 0 \\ 0 & 12EI/\ell^3 & 6EI/\ell^2 \\ 0 & 6EI/\ell^2 & 4EI/\ell \end{bmatrix} \tag{3.30}$$

$$[\kappa_{12}] = \begin{bmatrix} -AE/\ell & 0 & 0 \\ 0 & -12EI/\ell^3 & 6EI/\ell^2 \\ 0 & -6EI/\ell^2 & 2EI/\ell \end{bmatrix} \tag{3.31}$$

$$[\kappa_{21}] = \begin{bmatrix} -AE/\ell & 0 & 0 \\ 0 & -12EI/\ell^3 & -6EI/\ell^2 \\ 0 & 6EI/\ell^2 & 2EI/\ell \end{bmatrix} \tag{3.32}$$

$$[\kappa_{22}] = \begin{bmatrix} AE/\ell & 0 & 0 \\ 0 & 12EI/\ell^3 & -6EI/\ell^2 \\ 0 & -6EI/\ell^2 & 4EI/\ell \end{bmatrix} \tag{3.33}$$

Therefore, the element stiffness matrix is

$$[k] = \begin{bmatrix} \kappa_{11} & \kappa_{12} \\ \kappa_{21} & \kappa_{22} \end{bmatrix}$$

$$= \begin{bmatrix} AE/\ell & 0 & 0 & -AE/\ell & 0 & 0 \\ 0 & 12EI/\ell^3 & 6EI/\ell^2 & 0 & -12EI/\ell^3 & 6EI/\ell^2 \\ 0 & 6EI/\ell^2 & 4EI/\ell & 0 & -6EI/\ell^2 & 2EI/\ell \\ -AE/\ell & 0 & 0 & AE/\ell & 0 & 0 \\ 0 & -12EI/\ell^3 & -6EI/\ell^2 & 0 & 12EI/\ell^3 & -6EI/\ell^2 \\ 0 & 6EI/\ell^2 & 2EI/\ell & 0 & -6EI/\ell^2 & 4EI/\ell \end{bmatrix}$$

$$(3.34)$$

3.4 TRANSFORMATION TO GLOBAL DIRECTIONS

Equation (3.29) may be written in the succinct form of Equation (3.6) as

$$\{f\} = [k]\{u\} \qquad (3.35)$$

This equation is directly analogous to Equation (2.8) for trusses. Recall, however, that the directions of the components of $\{f\}$ and $\{u\}$ of Equation (2.8) and also of Equation (3.35) are parallel and perpendicular to the element itself (that is, they are along the *local* or element directions). Hence, if the element is inclined at an angle γ relative to the horizontal and vertical directions, as in Figure 3.12, the components of $\{f\}$ and $\{u\}$ of Equation (3.35) will no longer be parallel to the horizontal and vertical directions. Since we are ultimately concerned with force and displacement components in the horizontal and vertical directions (that is, the *global* directions), it is necessary to transform the components of $\{f\}$ and $\{u\}$ into the global directions. Therefore, the objective of this section is to develop this transformation.

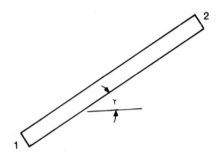

Figure 3.12   Inclined Beam Element

We can obtain this component transformation by following the same procedures as developed in Section 3.4 for trusses: Specifically, let Equation (3.34) be rewritten in the form

$$\{f'\} = [k'] \{u'\} \tag{3.36}$$

where the primes indicate quantities referred to the *local* or element directions. Using this notation, Equation (3.29), the partitioned form of Equation (3.36), becomes

$$\begin{bmatrix} f_1' \\ -- \\ f_2' \end{bmatrix} = \begin{bmatrix} \kappa_{11}' & \mid & \kappa_{12}' \\ --+-- \\ \kappa_{21}' & \mid & \kappa_{22}' \end{bmatrix} \begin{bmatrix} u_1' \\ -- \\ u_2' \end{bmatrix} \tag{3.37}$$

Next, let unprimed quantities be referred to *global* directions and consider the displacement components of end 1 of the element. These are represented in Figure 3.13.

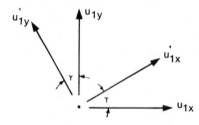

Figure 3.13   Displacements of End 1 Along
Local and Global Directions

From this sketch we see that

$$u_{1x} = u'_{1x}c - u'_{1y}s$$
$$u_{1y} = u'_{1x}s + u'_{1y}c \qquad (3.38)$$
$$\theta_1 = \theta'_1$$

where, as before, s and c are abbreviations for $\sin \gamma$ and $\cos \gamma$. In matrix form, Equations (3.38) may be written

$$\begin{bmatrix} u_{1x} \\ u_{1y} \\ \theta_1 \end{bmatrix} = \begin{bmatrix} c & -s & 0 \\ s & c & 0 \\ 0 & 0 & 1 \end{bmatrix} \begin{bmatrix} u'_{1x} \\ u'_{1y} \\ \theta'_1 \end{bmatrix} \qquad (3.39)$$

or

$$\{u_1\} = [t]\{u'_1\} \qquad (3.40)$$

where $[t]$ is defined by comparison of the expressions.

In a directly analogous manner, we see that at end 2 of the element

$$\{u_2\} = [t]\{u'_2\} \qquad (3.41)$$

where $[t]$ is the same matrix as in Equation (3.40). Similarly, for the force arrays, we have

$$\{f_1\} = [t]\{f'_1\} \qquad (3.42)$$

and

$$\{f_2\} = [t]\{f'_2\} \qquad (3.43)$$

Equations (3.40) and (3.41) may be consolidated in the form

$$\{u\} = \begin{bmatrix} u_1 \\ -- \\ u_2 \end{bmatrix} = \begin{bmatrix} t & | & 0 \\ --|-- \\ 0 & | & t \end{bmatrix} \begin{bmatrix} u'_1 \\ -- \\ u'_2 \end{bmatrix} = [T]\{u'\} \qquad (3.44)$$

where $[T]$ is defined by inspection. Similarly,

$$\{f\} = [T]\{f'\} \qquad (3.45)$$

It is readily seen that $[T]$ is orthogonal. Hence, Equations (3.44) and (3.45) may be solved for $\{u'\}$ and $\{f'\}$ as

$$\{u'\} = [T]^T \{u\}$$

$$\{f'\} = [T]^T \{f\}$$
(3.46)

Using Equations (3.45) and (3.46), Equation (3.36) may be written

$$[T] \{f'\} = \{f\} = [T][k'] \{u'\} = [T][k'][T]^T \{u\} \quad (3.47)$$

or simply as

$$\{f\} = [k] \{u\}$$
(3.48)

where

$$[k] = [T][k'][T]^T$$
(3.49)

By performing the indicated matrix multiplication of Equation (3.49), the "transformed" striffness matrix $[k]$ becomes as shown in Equation (3.50) [see Equation (3.33)].

Equation (3.50) is thus the desired force-displacement relation referred to the global directions.

On occasion, it is desirable to generalize this result to include the situation where the global direction would be *different* at the respective ends of the beam element. Such a generalization is outlined in Problems 3.7 and 3.8.

$$[k] = \frac{EI}{\ell^3}
\begin{bmatrix}
\left(\dfrac{A\ell^2}{I}c^2 + 12s^2\right) & \left(\dfrac{A\ell^2}{I}sc - 12sc\right) & (-6\ell s) & \left(-\dfrac{A\ell^2}{I}c^2 - 12s^2\right) & \left(-\dfrac{A\ell^2}{I}sc + 12sc\right) & (-6\ell s) \\[2ex]
\left(\dfrac{A\ell^2}{I}sc - 12sc\right) & \left(\dfrac{A\ell^2}{I}s^2 + 12c^2\right) & (6\ell c) & \left(-\dfrac{A\ell^2}{I}sc + 12sc\right) & \left(-\dfrac{A\ell^2}{I}s^2 - 12c^2\right) & (6\ell c) \\[2ex]
(-6\ell s) & (6\ell c) & (4\ell^2) & (6\ell s) & (-6\ell c) & (2\ell^2) \\[2ex]
\left(-\dfrac{A\ell^2}{I}c^2 - 12s^2\right) & \left(-\dfrac{A\ell^2}{I}sc + 12sc\right) & (6\ell s) & \left(\dfrac{A\ell^2}{I}c^2 + 12s^2\right) & \left(\dfrac{A\ell^2}{I}sc - 12sc\right) & (6\ell s) \\[2ex]
\left(-\dfrac{A\ell^2}{I}sc + 12sc\right) & \left(-\dfrac{A\ell^2}{I}s^2 - 12c^2\right) & (-6\ell c) & \left(\dfrac{A\ell^2}{I}sc - 12sc\right) & \left(\dfrac{A\ell^2}{I}s^2 + 12c^2\right) & (-6\ell c) \\[2ex]
(-6\ell s) & (6\ell c) & (2\ell^2) & (6\ell s) & (-6\ell c) & (4\ell^2)
\end{bmatrix}$$

$$(3.50)$$

3.5 APPLICATION WITH A SIMPLE EXAMPLE

Before we consider in detail the solution of our primary example problem of Figure 3.1, it is instructive to illustrate assembly and solution procedures with a few simple examples. As mentioned earlier, the assembly procedure for beam elements is virtually the same as the assembly procedure developed in detail for trusses. Therefore, we will not repeat that development here, but instead we will simply illustrate its application with beam elements.

As a first example, consider a beam with fixed ends and loaded with a force P at its center as shown in Figure 3.14. Let the beam be divided into two equal length beam elements as shown in Figure 3.15. The beam thus has three joints or nodes. Since there are three force components and three displacement components at each node, the global force and displacement arrays will each have 9 components or elements. The global stiffness matrix will have 81 elements. These global arrays may be obtained from the local arrays by using incidence matrices as described in Section 3.6. Specifically,

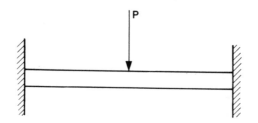

Figure 3.14 Centrally Loaded, Fixed End Beam

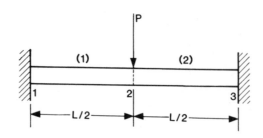

Figure 3.15 Beam Elements and Nodes for Fixed End Beam

for the beam of Figure 3.15, the global and local displacement arrays are related as (see Problem 3.9)

$$\{\overset{(e)}{U}\} = [\overset{(e)}{\Lambda}]^T \{\overset{(e)}{u}\} \tag{3.51}$$

and

$$\{\overset{(e)}{u}\} = [\overset{(e)}{\Lambda}]\{\overset{(e)}{U}\} = [\overset{(e)}{\Lambda}]\{U\} \tag{3.52}$$

where $e = 1, 2$ and where $[\overset{(1)}{\Lambda}]$ and $[\overset{(2)}{\Lambda}]$ are

$$[\overset{(1)}{\Lambda}] = \begin{bmatrix} 1 & 0 & 0 & 0 & 0 & 0 & 0 & 0 & 0 \\ 0 & 1 & 0 & 0 & 0 & 0 & 0 & 0 & 0 \\ 0 & 0 & 1 & 0 & 0 & 0 & 0 & 0 & 0 \\ 0 & 0 & 0 & 1 & 0 & 0 & 0 & 0 & 0 \\ 0 & 0 & 0 & 0 & 1 & 0 & 0 & 0 & 0 \\ 0 & 0 & 0 & 0 & 0 & 1 & 0 & 0 & 0 \end{bmatrix} \tag{3.53}$$

and

$$[\overset{(2)}{\Lambda}] = \begin{bmatrix} 0 & 0 & 0 & 1 & 0 & 0 & 0 & 0 & 0 \\ 0 & 0 & 0 & 0 & 1 & 0 & 0 & 0 & 0 \\ 0 & 0 & 0 & 0 & 0 & 1 & 0 & 0 & 0 \\ 0 & 0 & 0 & 0 & 0 & 0 & 1 & 0 & 0 \\ 0 & 0 & 0 & 0 & 0 & 0 & 0 & 1 & 0 \\ 0 & 0 & 0 & 0 & 0 & 0 & 0 & 0 & 1 \end{bmatrix} \tag{3.54}$$

That is, specifically,

$$\{\overset{(1)}{U}\} = \begin{bmatrix} U_1 \\ U_2 \\ U_3 \\ U_4 \\ U_5 \\ U_6 \\ 0 \\ 0 \\ 0 \end{bmatrix} = \begin{bmatrix} U_{1x} \\ U_{1y} \\ \theta_1 \\ U_{2x} \\ U_{2y} \\ \theta_2 \\ 0 \\ 0 \\ 0 \end{bmatrix} = \begin{bmatrix} 1 & 0 & 0 & 0 & 0 & 0 \\ 0 & 1 & 0 & 0 & 0 & 0 \\ 0 & 0 & 1 & 0 & 0 & 0 \\ 0 & 0 & 0 & 1 & 0 & 0 \\ 0 & 0 & 0 & 0 & 1 & 0 \\ 0 & 0 & 0 & 0 & 0 & 1 \\ 0 & 0 & 0 & 0 & 0 & 0 \\ 0 & 0 & 0 & 0 & 0 & 0 \\ 0 & 0 & 0 & 0 & 0 & 0 \end{bmatrix} \begin{bmatrix} \overset{(1)}{u_{1x}} \\ \overset{(1)}{u_{1y}} \\ \overset{(1)}{\theta_1} \\ \overset{(1)}{u_{2x}} \\ \overset{(1)}{u_{2y}} \\ \overset{(1)}{\theta_2} \end{bmatrix} = \begin{bmatrix} \overset{(1)}{U_{1x}} \\ \overset{(1)}{U_{1y}} \\ \overset{(1)}{\theta_1} \\ \overset{(1)}{U_{2x}} \\ \overset{(1)}{U_{2y}} \\ \overset{(1)}{\theta_2} \\ 0 \\ 0 \\ 0 \end{bmatrix}$$

$$(3.55)$$

and

$$\{\overset{(2)}{U}\} = \begin{bmatrix} 0 \\ 0 \\ 0 \\ U_4 \\ U_5 \\ U_6 \\ U_7 \\ U_8 \\ U_9 \end{bmatrix} = \begin{bmatrix} 0 \\ 0 \\ 0 \\ U_{2x} \\ U_{2y} \\ \theta_2 \\ U_{3x} \\ U_{3y} \\ \theta_3 \end{bmatrix} = \begin{bmatrix} 0 & 0 & 0 & 0 & 0 & 0 \\ 0 & 0 & 0 & 0 & 0 & 0 \\ 0 & 0 & 0 & 0 & 0 & 0 \\ 1 & 0 & 0 & 0 & 0 & 0 \\ 0 & 1 & 0 & 0 & 0 & 0 \\ 0 & 0 & 1 & 0 & 0 & 0 \\ 0 & 0 & 0 & 1 & 0 & 0 \\ 0 & 0 & 0 & 0 & 1 & 0 \\ 0 & 0 & 0 & 0 & 0 & 1 \end{bmatrix} \begin{bmatrix} \overset{(2)}{u_{1x}} \\ \overset{(2)}{u_{1y}} \\ \overset{(2)}{\theta_1} \\ \overset{(2)}{u_{2x}} \\ \overset{(2)}{u_{2y}} \\ \overset{(2)}{\theta_2} \end{bmatrix} = \begin{bmatrix} 0 \\ 0 \\ 0 \\ \overset{(2)}{U_{1x}} \\ \overset{(2)}{U_{1y}} \\ \overset{(2)}{\theta_1} \\ \overset{(2)}{U_{2x}} \\ \overset{(2)}{U_{2y}} \\ \overset{(2)}{\theta_2} \end{bmatrix}$$

$$(3.56)$$

$$
\begin{bmatrix}
u_{1x}^{(1)} \\
u_{1y}^{(1)} \\
\theta_{1}^{(1)} \\
u_{2x}^{(1)} \\
u_{2y}^{(1)} \\
\theta_{2}^{(1)}
\end{bmatrix}
=
\begin{bmatrix}
1 & 0 & 0 & 0 & 0 & 0 & 0 & 0 & 0 \\
0 & 1 & 0 & 0 & 0 & 0 & 0 & 0 & 0 \\
0 & 0 & 1 & 0 & 0 & 0 & 0 & 0 & 0 \\
0 & 0 & 0 & 1 & 0 & 0 & 0 & 0 & 0 \\
0 & 0 & 0 & 0 & 1 & 0 & 0 & 0 & 0 \\
0 & 0 & 0 & 0 & 0 & 1 & 0 & 0 & 0
\end{bmatrix}
\begin{bmatrix}
U_1 \\
U_2 \\
U_3 \\
U_4 \\
U_5 \\
U_6 \\
U_7 \\
U_8 \\
U_9
\end{bmatrix}
\tag{3.57}
$$

and

$$
\begin{bmatrix}
u_{1x}^{(2)} \\
u_{1y}^{(2)} \\
\theta_{1}^{(2)} \\
u_{2x}^{(2)} \\
u_{2y}^{(2)} \\
\theta_{2}^{(2)}
\end{bmatrix}
=
\begin{bmatrix}
0 & 0 & 0 & 1 & 0 & 0 & 0 & 0 & 0 \\
0 & 0 & 0 & 0 & 1 & 0 & 0 & 0 & 0 \\
0 & 0 & 0 & 0 & 0 & 1 & 0 & 0 & 0 \\
0 & 0 & 0 & 0 & 0 & 0 & 1 & 0 & 0 \\
0 & 0 & 0 & 0 & 0 & 0 & 0 & 1 & 0 \\
0 & 0 & 0 & 0 & 0 & 0 & 0 & 0 & 1
\end{bmatrix}
\begin{bmatrix}
U_1 \\
U_2 \\
U_3 \\
U_4 \\
U_5 \\
U_6 \\
U_7 \\
U_8 \\
U_9
\end{bmatrix}
\tag{3.58}
$$

Similarly, the global element force arrays for the beam of Figure 3.15 are related as

$$\{ \overset{(e)}{F} \} = [\overset{(e)}{\Lambda}]^T \{ \overset{(e)}{f} \} \tag{3.59}$$

where $e = 1, 2$. Specifically,

$$\{ \overset{(1)}{F} \} = \begin{bmatrix} F_{1x}^{(1)} \\ F_{1y}^{(1)} \\ M_1^{(1)} \\ F_{2x}^{(1)} \\ F_{2y}^{(1)} \\ M_2^{(1)} \\ 0 \\ 0 \\ 0 \end{bmatrix} = \begin{bmatrix} 1 & 0 & 0 & 0 & 0 & 0 \\ 0 & 1 & 0 & 0 & 0 & 0 \\ 0 & 0 & 1 & 0 & 0 & 0 \\ 0 & 0 & 0 & 1 & 0 & 0 \\ 0 & 0 & 0 & 0 & 1 & 0 \\ 0 & 0 & 0 & 0 & 0 & 1 \\ 0 & 0 & 0 & 0 & 0 & 0 \\ 0 & 0 & 0 & 0 & 0 & 0 \\ 0 & 0 & 0 & 0 & 0 & 0 \end{bmatrix} \begin{bmatrix} f_{1x}^{(1)} \\ f_{1y}^{(1)} \\ m_1^{(1)} \\ f_{2x}^{(2)} \\ f_{2y}^{(1)} \\ m_2^{(1)} \end{bmatrix} \tag{3.60}$$

and

$$\{ \overset{(2)}{F} \} = \begin{bmatrix} 0 \\ 0 \\ 0 \\ F_{1x}^{(2)} \\ F_{1y}^{(2)} \\ M_1^{(2)} \\ F_{2x}^{(2)} \\ F_{2y}^{(2)} \\ M_2^{(2)} \end{bmatrix} = \begin{bmatrix} 0 & 0 & 0 & 0 & 0 & 0 \\ 0 & 0 & 0 & 0 & 0 & 0 \\ 0 & 0 & 0 & 0 & 0 & 0 \\ 1 & 0 & 0 & 0 & 0 & 0 \\ 0 & 1 & 0 & 0 & 0 & 0 \\ 0 & 0 & 1 & 0 & 0 & 0 \\ 0 & 0 & 0 & 1 & 0 & 0 \\ 0 & 0 & 0 & 0 & 1 & 0 \\ 0 & 0 & 0 & 0 & 0 & 1 \end{bmatrix} \begin{bmatrix} f_{1x}^{(2)} \\ f_{1y}^{(2)} \\ m_1^{(2)} \\ f_{2x}^{(2)} \\ f_{2y}^{(2)} \\ m_2^{(2)} \end{bmatrix} \tag{3.61}$$

Note that a relation analogous to Equation (3.52) does not hold for forces; that is,

$$\{ \overset{(e)}{f} \} \neq [\overset{(e)}{\Lambda}]\{F\} \tag{3.62}$$

This is due to the fact that displacements at a node are common or equal at each adjoining element whereas the forces are different for different adjoining elements, depending upon node equilibrium considerations.

Recall that the fundamental force-displacement relation for an element in local form is

$$\{ \overset{(e)}{f} \} = [\overset{(e)}{k}]\{ \overset{(e)}{u} \} \tag{3.63}$$

The global form of the element force-displacement relation is then obtained from Equations (3.51), (3.52), and (3.63) as

$$\{ \overset{(e)}{F} \} = [\overset{(e)}{K}]\{ U \} \tag{3.64}$$

where the global element stiffness matrix $[\overset{(e)}{K}]$ is related to $[\overset{(e)}{k}]$ as

$$[\overset{(e)}{K}] = [\overset{(e)}{\Lambda}]^T [\overset{(e)}{k}] [\overset{(e)}{\Lambda}] \tag{3.65}$$

By using Equations (3.34), (3.53), and (3.54), the global element stiffness matrices $[\overset{(1)}{K}]$ and $[\overset{(2)}{K}]$ become

$$[K]^{(1)} = \begin{bmatrix}
2AE/L & 0 & 0 & -2AE/L & 0 & 0 & 0 & 0 \\
0 & 96EI/L^3 & 24EI/L^2 & 0 & -96EI/L^3 & 24EI/L^2 & 0 & 0 \\
0 & 24EI/L^2 & 8EI/L & 0 & -24EI/L^2 & 4EI/L & 0 & 0 \\
-2AE/L & 0 & 0 & 2AE/L & 0 & 0 & 0 & 0 \\
0 & -96EI/L^3 & -24EI/L^2 & 0 & 96EI/L^3 & -24EI/L^2 & 0 & 0 \\
0 & 24EI/L^2 & 4EI/L & 0 & -24EI/L^2 & 8EI/L & 0 & 0 \\
0 & 0 & 0 & 0 & 0 & 0 & 0 & 0 \\
0 & 0 & 0 & 0 & 0 & 0 & 0 & 0
\end{bmatrix} \qquad (3.66)$$

$$[K]^{(2)} = \begin{bmatrix}
0 & 0 & 0 & 0 & 0 & 0 & 0 & 0 \\
0 & 0 & 0 & 0 & 0 & 0 & 0 & 0 \\
0 & 0 & 2AE/L & 0 & 0 & -2AE/L & 0 & 0 \\
0 & 0 & 0 & 96EI/L^3 & 24EI/L^2 & 0 & -96EI/L^3 & 24EI/L^2 \\
0 & 0 & 0 & 24EI/L^2 & 8EI/L & 0 & -24EI/L^2 & 4EI/L \\
0 & 0 & -2AE/L & 0 & 0 & 2AE/L & 0 & 0 \\
0 & 0 & 0 & -96EI/L^3 & -24EI/L^2 & 0 & 96EI/L^3 & -24EI/L^2 \\
0 & 0 & 0 & 24EI/L^2 & 4EI/L & 0 & -24EI/L^2 & 8EI/L
\end{bmatrix} \qquad (3.67)$$

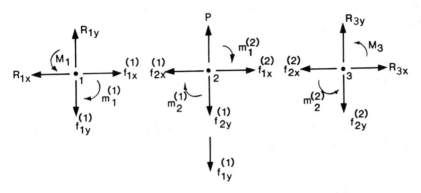

Figure 3.16 Free Body Diagrams of the Nodes

As in Section 2.6, let us now consider the equilibrium of the nodes of the beam of Figure 3.15. Free body diagrams of the three nodes are shown in Figure 3.16, where $R_{1x}, R_{1y}, M_1, R_{3x}, R_{3y}$, and M_3 represent the reaction forces and moments at the fixed supports. The equilibrium considerations immediately lead to the following relations:

$$\begin{bmatrix} f_{1x}^{(1)} \\ f_{1y}^{(1)} \\ m_1^{(1)} \end{bmatrix} = \begin{bmatrix} R_{1x} \\ R_{1y} \\ M_1 \end{bmatrix} \tag{3.68}$$

$$\begin{bmatrix} f_{2x}^{(1)} + f_{1x}^{(2)} \\ f_{2y}^{(1)} + f_{1y}^{(2)} \\ m_2^{(1)} + m_1^{(2)} \end{bmatrix} = \begin{bmatrix} 0 \\ -P \\ 0 \end{bmatrix} \tag{3.69}$$

and

$$\begin{bmatrix} f_{2x}^{(2)} \\ f_{2y}^{(2)} \\ m_2^{(2)} \end{bmatrix} = \begin{bmatrix} R_{3x} \\ R_{3y} \\ M_3 \end{bmatrix} \tag{3.70}$$

By examining Equations (3.60) and (3.61), we see that these relations may be written

$$\sum_{e=1}^{2} \{\overset{(e)}{F}\} = \{\overset{(1)}{F}\} + \{\overset{(2)}{F}\} = \{P\} = \{L\} + \{R\} \tag{3.71}$$

where $\{L\}$ and $\{R\}$ are the externally applied load and support reaction arrays, respectively, and are given by

$$\{L\} = \begin{bmatrix} 0 \\ 0 \\ 0 \\ 0 \\ -P \\ 0 \\ 0 \\ 0 \\ 0 \end{bmatrix} \qquad \{R\} = \begin{bmatrix} R_{1x} \\ R_{1y} \\ M_1 \\ 0 \\ 0 \\ 0 \\ R_{3x} \\ R_{3y} \\ M_3 \end{bmatrix} \tag{3.72}$$

By substituting from Equation (3.64), Equation (3.71) may be written

$$\{L\} + \{R\} = \sum_{e=1}^{2} [\overset{(e)}{K}]\{U\} = ([\overset{(1)}{K}] + [\overset{2}{K}])\{U\} = [K]\{U\} \tag{3.73}$$

where $[K]$, the global stiffness matrix for the system is defined as the sum of the global element stiffness matrices. From Equations (3.66) and (3.67), it is seen that $[K]$ is

$$[K] = \begin{bmatrix}
2AE/L & 0 & 0 & -2AE/L & 0 & 0 & 0 & 0 & 0 \\
0 & 96EI/L^3 & 24EI/L^2 & 0 & -96EI/L^3 & 24EI/L^2 & 0 & 0 & 0 \\
0 & 24EI/L^2 & 8EI/L & 0 & -24EI/L^2 & 4EI/L & 0 & 0 & 0 \\
-2AE/L & 0 & 0 & 4AE/L & 0 & 0 & -2AE/L & 0 & 0 \\
0 & -96EI/L^3 & -24EI/L^2 & 0 & 192EI/L^3 & 0 & 0 & -96EI/L^3 & 24EI/L^2 \\
0 & 24EI/L^2 & 4EI/L & 0 & 0 & 16EI/L & 0 & -24EI/L^2 & 4EI/L \\
0 & 0 & 0 & -2AE/L & 0 & 0 & 2AE/L & 0 & 0 \\
0 & 0 & 0 & 0 & -96EI/L^3 & -24EI/L^2 & 0 & 96EI/L^3 & -24EI/L^2 \\
0 & 0 & 0 & 0 & 24EI/L^2 & 4EI/L & 0 & -24EI/L^2 & 8EI/L
\end{bmatrix} \qquad (3.74)$$

By substituting Equations (3.72) and (3.74) into the governing equation (3.73) and by enforcing the end conditions

$$U_1 = U_2 = U_3 = U_7 = U_8 = U_9 = 0 \qquad (3.75)$$

the resulting equations may be solved for the unknown support reactions and the displacements at node 2. The results (see Problem 3.10) are

$$
\begin{aligned}
R_{1x} &= R_{3x} = 0 \\
R_{1y} &= R_{3y} = P/2 \\
M_1 &= -M_3 = PL/8 \qquad\qquad (3.76) \\
U_4 &= U_6 = 0 \\
U_5 &= -PL^3/192EI
\end{aligned}
$$

3.6   FIXED END LOADS: DISTRIBUTED LOADING

The foregoing example illustrates the assembly and solution procedure for a fixed-end beam with a single concentrated load. However, we are generally interested in solving problems with more elaborate loading conditions. For example, it would be helpful to have a procedure for finding the unknown displacements and reaction forces for a beam with an arbitrary loading function $P(x)$ such as shown in Figure 3.17.

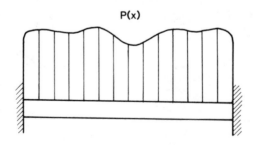

Figure 3.17   Fixed-End Beam with Arbitrary Loading Function

The procedure of Section 3.5 does not address the problem of modeling such a loading function. Indeed, the finite element procedure we have developed, requires that the loads be applied at the element joints or nodes. This suggests that if we are to consider a loading such as in Figure 3.17, we should replace the loading by a series of concentrated forces which are approximately statically equivalent to the loading function $P(x)$. Figure 3.18 shows an example of such a replacement. The beam is thus divided into as many elements as are needed to obtain the desired accuracy in the loading approximation.

Although this approach appears to be simple, relatively "straightforward," and sufficiently accurate, at least, in a limiting sense, it is nevertheless still an approximation. But, a more serious difficulty of this approach is that it tends to introduce a large number of beam elements leading to an excessive computational burden.

Our objective in this section is to introduce a load modelling procedure which overcomes this difficulty. The procedure we are about to introduce has the advantage of replacing a given loading with a simpler statically equivalent loading but without the disadvantage of introducing a large number of beam elements.

The basic steps of this procedure are as follows:

1. Divide the beam into elements with joints at the supports, at the discontinuities, and at major changes in the loading functions, and at otherwise convenient positions.

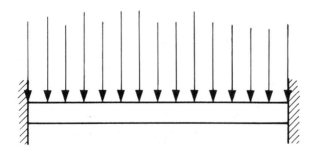

Figure 3.18 Approximate Statically Equivalent Loading
to $P(x)$ of Figure 3.17

2. Replace the loading over a typical beam element with loads at its joints which are exactly negative the support reactions of a fixed-end beam which has the same length as the element with the same loading as along the element.

For example, for the arbitrarily loaded beam of Figure 3.19, let us introduce three beam elements as shown in Figure 3.20. Consider, for example, element (2): The corresponding fixed-end beam with the same length and loading as element (2) is shown in Figure 3.21. The negative of the support reactions of this beam are depicted in Figure 3.22. This is the joint loading to be used at the joints of element (2) of the beam of Figure 3.20.

To illustrate this procedure with a simple specific example, consider the uniformly loaded, fixed-end beam of length L shown in Figure 3.23 where P_0 is the load intensity per unit beam length.

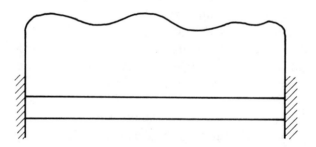

Figure 3.19   Fixed-End Beam with
Arbitrary Loading Function

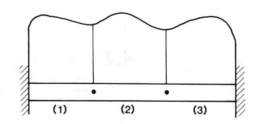

Figure 3.20   Arbitrarily Loaded Beam Divided
into Three Elements

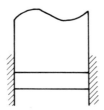

Figure 3.21   Element (2) of Figure 3.20
as a Fixed-End Beam

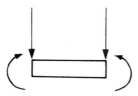

Figure 3.22   Negative of the Support Reactions
of the Beam of Figure 3.21

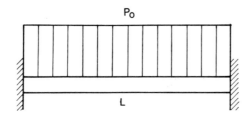

Figure 3.23   Uniformly Loaded, Fixed-End Beam

It is easily shown (see Problem 3.12) that the support reactions for this beam are as shown in Figure 3.24 where R_1, R_2, M_1, and M_2 are

$$R_1 = R_2 = P_0 L/2$$

$$M_1 = -M_2 = P_0 L^2/12$$

(3.77)

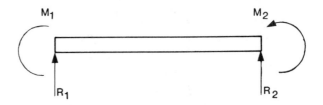

Figure 3.24 Support Reactions for the
Fixed-End Beam of Figure 3.23

Now, following the procedure outlined above, let us make an analysis of a simply supported, uniformly loaded beam of length L as shown in Figure 3.25, where P_0 is the load intensity per unit beam length. Specifically, let the beam be divided into two equal length elements as shown in Figure 3.26 where the support reactions are also shown.

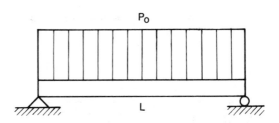

Figure 3.25 Uniformly Loaded, Simply Supported Beam

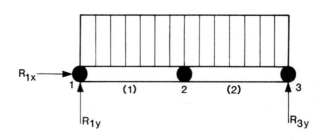

Figure 3.26 Finite Element Model of the Beam of Figure 3.25

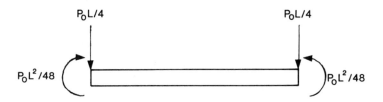

Figure 3.27   Equivalent Superposed Element Joint
Loading for the Beam of Figure 3.26

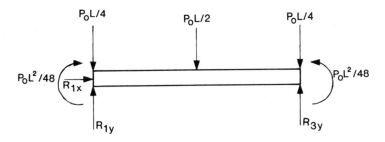

Figure 3.28   Equivalent Joint Loading for the Beam
of Figures 3.25 and 3.26

Next, let the loading on each element be replaced by loadings, superposed on the joints, which are exactly negative the support reactions
of fixed-end beams with the same length and loading as the elements.
Hence, for the beam elements of Figure 3.25, the equivalent superposed loading is as shown in Figure 3.27. [The magnitudes of the
forces and moments are obtained directly from Equation (3.77) by
replacing L by $L/2$.] Therefore, the equivalent joint loading for the
entire beam is as shown in Figure 3.28. The global stiffness matrix
$[K]$ for this beam is listed in Equation (3.74). From Figures 3.25,
3.26 and 3.28, the global displacement, loading and reaction arrays
are

$$\{U\} = \begin{bmatrix} 0 \\ 0 \\ \theta_1 \\ U_{2x} \\ U_{2y} \\ \theta_2 \\ U_{3x} \\ 0 \\ \theta_3 \end{bmatrix} \qquad \{L\} = \begin{bmatrix} 0 \\ -P_0 L/4 \\ -P_0 L^2/48 \\ 0 \\ -P_0 L/2 \\ 0 \\ 0 \\ -P_0 L/4 \\ P_0 L^2/48 \end{bmatrix} \qquad \{R\} = \begin{bmatrix} R_{1x} \\ R_{1y} \\ 0 \\ 0 \\ 0 \\ 0 \\ 0 \\ R_{3y} \\ 0 \end{bmatrix} \qquad (3.78)$$

Hence, the governing matrix equation is

$$\{L\} + \{R\} = [K]\{U\} \qquad (3.79)$$

Using the expression for $[K]$ of Equation (3.74), Equation (3.79) is easily shown by matrix multiplication to be equivalent to the following nine scalar equations:

$$R_{1x} = -(2AE/L)\,U_{2x}$$

$$R_{1y} - (P_0/4) = -(96EI/L^3)\,U_{2y} + (24EI/L^2)\,\theta_1 + (24EI/L^2)\,\theta_2$$

$$-(P_0 L^2/48) = -(24EI/L^2)\,U_{2y} + (8EI/L)\,\theta_1 + (4EI/L)\,\theta_2$$

$$0 = (4AE/L)\,U_{2x} - (2AE/L)\,U_{3x}$$

$$-P_0 L/2 = (192EI/L^3)\,U_{2y} - (24EI/L^2)\,\theta_1 + (24EI/L^2)\,\theta_3$$
$$(3.80)$$

$$0 = (16EI/L)\,\theta_2 + (4EI/L)\,\theta_1 + (4EI/L)\,\theta_3$$

$$0 = -(2AE/L)\,U_{2x} + (2AE/L)\,U_{3x}$$

$$R_{3y} - P_0/4 = -(96EI/L^3)\,U_{2y} - (24EI/L^2)\,\theta_2 - (24EI/L^2)\,\theta_3$$

$$(P_0 L/48) = (24EI/L^2)\,U_{2y} + (4EI/L)\,\theta_2 + (8EI/L)\,\theta_3$$

Solving these equations for the unknown displacements and reactions leads to the results

$$\theta_1 = P_0 L^3/24EI \qquad\qquad \theta_3 = P_0 L^3/24EI$$

$$U_{2x} = 0 \qquad\qquad R_{1x} = 0$$

$$U_{2y} = -5P_0 L^4/384EI \qquad\qquad R_{1y} = P_0 L/2 \qquad (3.81)$$

$$\theta_2 = 0 \qquad\qquad R_{3y} = P_0 L/2$$

$$U_{3x} = 0$$

3.7 JOINT RELEASES

Recall that our objective in this chapter is to develop procedures for analyzing general beam and frame systems. Our approach in meeting this objective is to study various aspects of beam and frame loading and support systems and then to illustrate the finite element procedures for these situations through a series of simple examples.

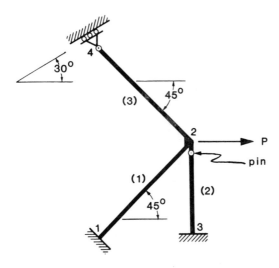

Figure 3.29 Example Frame with Labeled
Elements and Joints

Recall further, that for our concluding example we plan to consider the frame system of Figure 3.1. Figure 3.29 shows that system again with the joints and elements labeled. Notice that at joint 2, the elements are not all rigidly connected. That is, element (2) is *pinned* to the joint. This means, of course, that the moment on element (2), at joint 2, is zero. But, more than this, it means that element (2) is not strictly a fixed-beam element such as we have discussed and studied in the foregoing sections. Hence, we need to adjust and generalize our analysis to accommodate elements such as this. This "adjustment" is called a "joint release" since the fixed or rigid joint at the element end is replaced by, or "released" to, a pinned, or possibly a sliding joint. Therefore, our objective in this section is to develop element stiffness matrices analogous to the stiffness matrix of Equation (3.34) of Section 3.3 for elements with modified or released joints.

One approach to this development is simply to follow the steps of Section 3.3 for the modified element. That is, suppose an element has a pinned end (or "moment release") at its end 2. Its free-body diagram would then appear as shown in Figure 3.30. The force-displacement relations would then take the form

$$\{\hat{f}\} = [\hat{k}]\{\hat{u}\} \tag{3.82}$$

where $[\hat{k}]$ is a 5×5 element stiffness matrix and $\{\hat{f}\}$ and $\{\hat{u}\}$ are

$$\{\hat{f}\} = \begin{bmatrix} f_{1x} \\ f_{1y} \\ m_1 \\ f_{2x} \\ f_{2y} \end{bmatrix} \quad \text{and} \quad \{\hat{u}\} = \begin{bmatrix} u_{1x} \\ u_{1y} \\ \theta_1 \\ u_{2x} \\ u_{2y} \end{bmatrix} \tag{3.83}$$

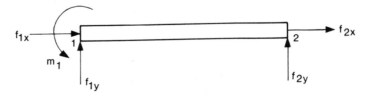

Figure 3.30   Free Body Diagram of a Beam Element
with a Moment Release at End 2

However, the desired element stiffness matrix $[k]_6$ of a two-dimensional beam element, with (or without) a joint release, should be 6 × 6.* That is, it should contain 6 rows and 6 columns so that it may be conveniently superposed with the usual beam element stiffness matrices (such as in Equation (3.34) of Section 3.3), in an assembly procedure. $[k]_6$ may be conveniently obtained from the 5 × 5 stiffness matrix $[k]$ of Equation (3.82) by simply appending a 6th row and a 6th column of zeros. That is, in partitioned form, $[k]_6$ is

$$[k]_6 = \begin{bmatrix} \hat{k} & | & 0 \\ --&|&-- \\ 0 & | & 0 \end{bmatrix} \tag{3.84}$$

In using the steps of Section 3.3 to obtain $[\hat{k}]$, it is convenient to partition the terms of Equation (3.82) as follows [see Equation (3.29)]:

$$\begin{bmatrix} \hat{f}_1 \\ -- \\ \hat{f}_1 \end{bmatrix} = \begin{bmatrix} \hat{k}_{11} & | & \hat{k}_{12} \\ --&+&-- \\ \hat{k}_{21} & | & \hat{k}_{22} \end{bmatrix} \begin{bmatrix} \hat{u}_1 \\ -- \\ \hat{u}_2 \end{bmatrix} \tag{3.85}$$

where $\{\hat{f}_1\}, \{\hat{f}_2\}, \{\hat{u}_1\}$, and $\{\hat{u}_2\}$ are

$$\{\hat{f}_1\} = \begin{bmatrix} f_{1x} \\ f_{1y} \\ m_1 \end{bmatrix} \qquad \{\hat{f}_2\} = \begin{bmatrix} f_{2x} \\ f_{2y} \end{bmatrix} \tag{3.86}$$

and

$$\{\hat{u}_1\} = \begin{bmatrix} u_{1x} \\ u_{1y} \\ \theta_1 \end{bmatrix} \qquad \{\hat{u}_2\} = \begin{bmatrix} u_{2x} \\ u_{2y} \end{bmatrix} \tag{3.87}$$

---

*Regarding notation, the subscript 6 on $[k]_6$ means that there is a moment release at end 2, that is, a release, or zero, in the 6th entry of $\{f\}$.

Then, by using the procedure of Section 3.3, of alternately loading and supporting the beam element of Figure 3.30 at ends 1 and 2, the following expressions are obtained for $[\hat{\kappa}_{11}]$, $[\hat{\kappa}_{12}]$, $[\hat{\kappa}_{21}]$, and $[\hat{\kappa}_{22}]$ (see Problems 3.14 to 3.17):

$$[\hat{\kappa}_{11}] = \begin{bmatrix} AE/\ell & 0 & 0 \\ 0 & 3EI/\ell^3 & 3EI/\ell^2 \\ 0 & 3EI/\ell^2 & 3EI/\ell \end{bmatrix} \tag{3.88}$$

$$[\hat{\kappa}_{12}] = \begin{bmatrix} -AE/\ell & 0 \\ 0 & -3EI/\ell^3 \\ 0 & -3EI/\ell^2 \end{bmatrix} \tag{3.89}$$

$$[\hat{\kappa}_{21}] = \begin{bmatrix} -AE/\ell & 0 & 0 \\ 0 & -3EI/\ell^3 & -3EI/\ell^2 \end{bmatrix} \tag{3.90}$$

and

$$[\hat{\kappa}_{22}] = \begin{bmatrix} AE/\ell & 0 \\ 0 & 3EI/\ell^3 \end{bmatrix} \tag{3.91}$$

Hence $[\hat{k}]$ and $[k]_6$ become

$$[\hat{k}] = \begin{bmatrix} AE/\ell & 0 & 0 & -AE/\ell & 0 \\ 0 & 3EI/\ell^3 & 3EI/\ell^2 & 0 & -3EI/\ell^3 \\ 0 & 3EI/\ell^2 & 3EI/\ell & 0 & -3EI/\ell^2 \\ -AE/\ell & 0 & 0 & AE/\ell & 0 \\ 0 & -3EI/\ell^3 & -3EI/\ell^2 & 0 & 3EI/\ell^3 \end{bmatrix}$$

$$\tag{3.92}$$

and

$$[k]_6 = \begin{bmatrix} AE/\ell & 0 & 0 & -AE/\ell & 0 & 0 \\ 0 & 3EI/\ell^3 & 3EI/\ell^2 & 0 & -3EI/\ell^3 & 0 \\ 0 & 3EI/\ell^2 & 3EI/\ell & 0 & -3EI/\ell^2 & 0 \\ -AE/\ell & 0 & 0 & AE/\ell & 0 & 0 \\ 0 & -3EI/\ell^3 & -3EI/\ell^2 & 0 & 3EI/\ell^3 & 0 \\ 0 & 0 & 0 & 0 & 0 & 0 \end{bmatrix}$$

(3.93)

It is interesting and enlightening to observe that the stiffness matrices of Equations (3.92) and (3.93) may also be obtained through partitioning of the terms of the basic element force-displacement equation of a fixed beam element [see Equation (3.6) of Section 3.3], and by manipulating the partitioned elements into the form of matrix $[k]_6$ of Equation (3.84). That is, let Equation (3.6) be partitioned as follows:

$$\{f\} = \begin{bmatrix} f_1 \\ -- \\ f_6 \end{bmatrix} = \begin{bmatrix} \kappa_{11} & | & \kappa_{16} \\ --- & + & --- \\ \kappa_{61} & | & \kappa_{66} \end{bmatrix} \begin{bmatrix} u_1 \\ -- \\ u_6 \end{bmatrix} = [k]\{u\}$$

(3.94)

where $\{f_1\}$, $\{f_6\}$, $\{u_1\}$, and $\{u_6\}$ are identified from the following expressions:

$$\{f\} = \begin{bmatrix} f_1 \\ -- \\ f_6 \end{bmatrix} = \begin{bmatrix} f_{1x} \\ f_{1y} \\ m_1 \\ f_{2x} \\ f_{2y} \\ -- \\ m_2 \end{bmatrix} \quad \text{and} \quad \{u\} = \begin{bmatrix} u_1 \\ -- \\ u_6 \end{bmatrix} = \begin{bmatrix} u_{1x} \\ u_{1y} \\ \theta_1 \\ u_{2x} \\ u_{2y} \\ -- \\ \theta_2 \end{bmatrix}$$

(3.95)

$[\kappa_{11}]$, $[\kappa_{16}]$, $[\kappa_{61}]$, and $[\kappa_{66}]$ may be identified similarly by the partitioning of $[k]$ of Equation (3.34). That is,

$$[\kappa_{11}] = \begin{bmatrix} AE/\ell & 0 & 0 & -AE/\ell & 0 \\ 0 & 12EI/\ell^3 & 6EI/\ell^2 & 0 & -12EI/\ell^3 \\ 0 & 6EI/\ell^2 & 4EI/\ell & 0 & -6EI/\ell^2 \\ -AE/\ell & 0 & 0 & AE/\ell & 0 \\ 0 & -12EI/\ell^3 & -6EI/\ell^2 & 0 & 12EI/\ell^3 \end{bmatrix}$$

$$(3.96)$$

$$[\kappa_{16}] = \begin{bmatrix} 0 \\ 6EI/\ell^2 \\ 2EI/\ell \\ 0 \\ -6EI/\ell^2 \end{bmatrix} \qquad (3.97)$$

$$[\kappa_{61}] = [0 \quad 6EI/\ell^2 \quad 2EI/\ell \quad 0 \quad -6EI/\ell^2] \qquad (3.98)$$

and

$$[\kappa_{66}] = [4EI/\ell] \qquad (3.99)$$

Hence, by expanding the matrices of Equation (3.94), we obtain

$$[\kappa_{11}]\{u_1\} + [\kappa_{16}]\{u_6\} = \{f_1\} \qquad (3.100)$$

and

$$[\kappa_{61}]\{u_1\} + [\kappa_{66}]\{u_6\} = \{f_6\} \qquad (3.101)$$

Now, if end 2 has a moment release, that is, if m_2 is zero, then

$$\{f_6\} = 0 \qquad (3.102)$$

Then, by solving Equation (3.101) for u_6, we obtain

$$\{u_6\} = -[\kappa_{66}]^{-1} [\kappa_{61}]\{u_1\} \tag{3.103}$$

Substituting this expression for $\{u_6\}$ into Equation (3.100) leads to

$$\{f_1\} = [\kappa_{11}]\{u_1\} - [\kappa_{16}] [\kappa_{66}]^{-1} [\kappa_{61}]\{u_1\} \tag{3.104}$$

or

$$\{f_1\} = [\hat{k}]\{u_1\} \tag{3.105}$$

where $[\hat{k}]$ is given by

$$[\hat{k}] = [\kappa_{11}] - [\kappa_{16}] [\kappa_{66}]^{-1}[\kappa_{61}] \tag{3.106}$$

Using Equations (3.96) to (3.99), it is easily shown (see Problem 3.19) that this expression for $[\hat{k}]$ is exactly the same as the expression in Equation (3.92).

3.8 EXAMPLE: A SIMPLE FRAME

It is helpful to demonstrate the application of the foregoing ideas on joint releases, with a simple example. In the next section, we will consider the more general example of the beam-frame system shown in Figures 3.1 and 3.29.

Consider the simple frame shown in Figure 3.31 which consists of two beams cantilevered or fixed, in their supports, rigidly attached to each other, and loaded at the attach point with a horizontal force P as shown. The length of the beams are $\ell_1 = 5m$ and $\ell_2 = 4m$ as shown in Figure 3.31. Both beams have identical cross sections with area $A = 10^{-3} m^2$ and second moment of area $I = 2.5 \times 10^{-2} m^4$. The modulus of elasticity is $E = 2.07 \times 10^{11} N/m^2$ for each beam.

To make a finite element analysis of this simple frame, let it be divided into two beam elements with three joints as shown in Figure 3.32. Following the procedure of Section 3.3 and specifically, Equation (3.34), it is seen that the element stiffness matrix of beam element (1) is

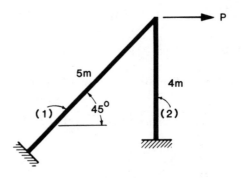

Figure 3.31 Simple Frame with Rigidly
Attached Cantilevered Beams

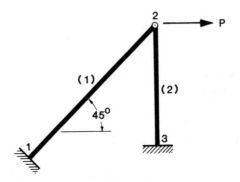

Figure 3.32 Elements and Joints of the
Simple Frame of Figure 3.31

$$
[\overset{(1)}{k}] = 10^8
\begin{bmatrix}
0.414 & 0.0 & 0.0 & -0.414 & 0.0 & 0.0 \\
0.0 & 4.968 & 12.42 & 0.0 & -4.968 & 12.42 \\
0.0 & 12.42 & 41.40 & 0.0 & -12.42 & 20.70 \\
-0.414 & 0.0 & 0.0 & 0.414 & 0.0 & 0.0 \\
0.0 & -4.968 & -12.42 & 0.0 & 4.968 & -12.42 \\
0.0 & 12.42 & 20.70 & 0.0 & -12.42 & 41.40
\end{bmatrix}
$$

$$(3.107)$$

where end 1 is at joint 1 and end 2 is at joint 2 and the units are N/m, N, and Nm depending upon the row and column. Using Equation (3.50) and transforming this to the global (horizontal and vertical directions), we obtain

$$\overset{(1)}{[k]} = 10^8 \begin{bmatrix} 2.691 & -2.277 & -8.782 & -2.691 & 2.277 & -8.782 \\ -2.277 & 2.691 & 8.782 & 2.277 & -2.691 & 8.782 \\ -8.782 & 8.782 & 41.40 & 8.782 & -8.782 & 20.70 \\ -2.691 & 2.277 & 8.782 & 2.691 & -2.277 & 8.782 \\ 2.277 & -2.691 & -8.782 & -2.277 & 2.691 & -8.782 \\ -8.782 & 8.782 & 20.70 & 8.782 & -8.782 & 41.40 \end{bmatrix}$$

$$(3.108)$$

Similarly, the stiffness matrix of beam element (2) referred to the global directions is

$$\overset{(2)}{[k]} = 10^8 \begin{bmatrix} 9.703 & 0.0 & -19.41 & -9.703 & 0.0 & -19.41 \\ 0.0 & 0.5175 & 0.0 & 0.0 & -0.5175 & 0.0 \\ -19.41 & 0.0 & 51.75 & 19.41 & 0.0 & 25.86 \\ -9.703 & 0.0 & 19.41 & 9.703 & 0.0 & 19.41 \\ 0.0 & -0.5175 & 0.0 & 0.0 & 0.5175 & 0.0 \\ -19.41 & 0.0 & 25.86 & 19.41 & 0.0 & 51.75 \end{bmatrix}$$

$$(3.109)$$

where end 1 is at joint 3 and end 2 is at joint 2.

For computational and conceptual purposes, it is convenient to partition these matrices as follows:

$$
\overset{(1)}{[k]} = \left[\begin{array}{c|c} \overset{(1)}{\kappa_{11}} & \overset{(1)}{\kappa_{12}} \\ \hline \overset{(1)}{\kappa_{21}} & \overset{(1)}{\kappa_{22}} \end{array}\right] \quad \text{and} \quad \overset{(2)}{[k]} = \left[\begin{array}{c|c} \overset{(2)}{\kappa_{11}} & \overset{(2)}{\kappa_{12}} \\ \hline \overset{(2)}{\kappa_{21}} & \overset{(2)}{\kappa_{22}} \end{array}\right] \quad (3.110)
$$

Hence, by inspection of Equations (3.108) and (3.109), the submatrices are

$$
[\overset{(1)}{\kappa_{11}}] = 10^8 \begin{bmatrix} 2.691 & -2.277 & -8.782 \\ -2.277 & 2.691 & 8.782 \\ -8.782 & 8.782 & 41.40 \end{bmatrix} \quad (3.111)
$$

$$
[\overset{(1)}{\kappa_{12}}] = [\overset{(1)}{\kappa_{21}}]^T = 10^8 \begin{bmatrix} -2.691 & 2.277 & -8.782 \\ 2.277 & -2.691 & 8.782 \\ 8.782 & -8.782 & 20.70 \end{bmatrix} \quad (3.112)
$$

$$
[\overset{(1)}{\kappa_{22}}] = 10^8 \begin{bmatrix} 2.691 & -2.277 & 8.782 \\ -2.277 & 2.691 & -8.782 \\ 8.782 & -8.782 & 41.40 \end{bmatrix} \quad (3.113)
$$

$$
[\overset{(2)}{\kappa_{11}}] = 10^8 \begin{bmatrix} 9.703 & 0.0 & -19.41 \\ 0.0 & 0.5175 & 0.0 \\ -19.41 & 0.0 & 51.75 \end{bmatrix} \quad (3.114)
$$

$$
[\overset{(2)}{\kappa_{12}}] = [\overset{(2)}{\kappa_{21}}]^T = 10^8 \begin{bmatrix} -9.703 & 0.0 & -19.41 \\ 0.0 & -0.5175 & 0.0 \\ 19.41 & 0.0 & 25.86 \end{bmatrix} \quad (3.115)
$$

$$[\overset{(2)}{\kappa_{22}}] = 10^8 \begin{bmatrix} 9.703 & 0.0 & 19.41 \\ 0.0 & 0.5175 & 0.0 \\ 19.41 & 0.0 & 51.75 \end{bmatrix} \quad (3.116)$$

Similarly, let the global stiffness matrix $[K]$ for the system be partitioned

$$[K] = \begin{bmatrix} K_{11} & K_{12} & K_{13} \\ K_{21} & K_{22} & K_{23} \\ K_{31} & K_{32} & K_{33} \end{bmatrix} \quad (3.117)$$

Then, by superposing the arrays of $[\overset{(1)}{k}]$ and $[\overset{(2)}{k}]$ and submatrices of $[K]$ become

$$[K_{11}] = [\overset{(1)}{\kappa_{11}}] = 10^8 \begin{bmatrix} 2.691 & -2.277 & -8.782 \\ -2.277 & 2.691 & 8.782 \\ -8.782 & 8.782 & 41.40 \end{bmatrix} \quad (3.118)$$

$$[K_{12}] = [K_{21}]^T = [\overset{(1)}{\kappa_{12}}] = 10^8 \begin{bmatrix} -2.691 & 2.277 & -8.782 \\ 2.277 & -2.691 & 8.782 \\ 8.782 & -8.782 & 20.70 \end{bmatrix} \quad (3.119)$$

$$[K_{13}] = [K_{31}]^T = \begin{bmatrix} 0.0 & 0.0 & 0.0 \\ 0.0 & 0.0 & 0.0 \\ 0.0 & 0.0 & 0.0 \end{bmatrix} \quad (3.120)$$

$$[K_{22}] = [\overset{(1)}{\kappa_{22}}] + [\overset{(2)}{\kappa_{22}}] = 10^8 \begin{bmatrix} 12.39 & -2.277 & 28.19 \\ -2.277 & 3.209 & -8.782 \\ 28.19 & -8.782 & 93.15 \end{bmatrix} \quad (3.121)$$

$$[K_{23}] = [K_{32}]^T = [\overset{(2)}{\kappa_{21}}] = 10^8 \begin{bmatrix} -9.703 & 0.0 & 19.41 \\ 0.0 & -0.5175 & 0.0 \\ -19.41 & 0.0 & 25.86 \end{bmatrix} \quad (3.122)$$

$$[K_{33}] = [\overset{(2)}{\kappa_{11}}] = 10^8 \begin{bmatrix} 2.691 & -2.277 & 8.782 \\ -2.277 & 2.691 & -8.782 \\ 8.782 & -8.782 & 41.40 \end{bmatrix} \quad (3.123)$$

Hence, $[K]$ itself becomes

$$[K] = 10^8 \begin{bmatrix}
2.691 & -2.277 & -8.782 & -2.691 & 2.277 & -8.782 & 0.0 & 0.0 & 0.0 \\
-2.277 & 2.691 & 8.782 & 2.277 & -2.691 & 8.782 & 0.0 & 0.0 & 0.0 \\
-8.782 & 8.782 & 41.40 & 8.782 & -8.782 & 20.70 & 0.0 & 0.0 & 0.0 \\
-2.691 & 2.277 & 8.782 & 12.39 & -2.277 & 28.19 & -9.703 & 0.0 & 19.41 \\
2.277 & -2.691 & -8.872 & -2.277 & 3.209 & -8.782 & 0.0 & -0.5175 & 0.0 \\
-8.782 & 8.782 & 20.70 & 28.19 & -8.782 & 93.15 & -19.41 & 0.0 & 25.87 \\
0.0 & 0.0 & 0.0 & -9.703 & 0.0 & -19.41 & 9.703 & 0.0 & -19.41 \\
0.0 & 0.0 & 0.0 & 0.0 & -0.5175 & 0.0 & 0.0 & 0.5175 & 0.0 \\
0.0 & 0.0 & 0.0 & 19.41 & 0.0 & 25.87 & -19.41 & 0.0 & 51.75
\end{bmatrix}$$

$$(3.124)$$

The global force-displacement relations then take the form

$$\{F\} = [K]\{U\} \tag{3.125}$$

where $\{F\}$ and $\{U\}$ may be written and partitioned

$$
\begin{bmatrix} R_{1x} \\ R_{1y} \\ M_1 \\ P \\ 0 \\ 0 \\ R_{3x} \\ R_{3y} \\ M_3 \end{bmatrix}
=
\begin{bmatrix} R_1 \\ \hline P \\ \hline R_3 \end{bmatrix}
\quad \text{and} \quad
\{U\} =
\begin{bmatrix} 0 \\ 0 \\ 0 \\ U_{2x} \\ U_{2y} \\ \theta_2 \\ 0 \\ 0 \\ 0 \end{bmatrix}
=
\begin{bmatrix} 0 \\ \hline U_2 \\ \hline 0 \end{bmatrix}
\tag{3.126}
$$

Then, Equation (3.125) may be written

$$
\begin{bmatrix} R_1 \\ \hline P \\ \hline R_3 \end{bmatrix}
=
\begin{bmatrix} K_{11} & K_{12} & K_{13} \\ \hline K_{21} & K_{22} & K_{13} \\ \hline K_{31} & K_{32} & K_{33} \end{bmatrix}
\begin{bmatrix} 0 \\ \hline U_2 \\ \hline 0 \end{bmatrix}
\tag{3.127}
$$

Hence, by block matrix multiplication

$$\{R_1\} = [K_{12}]\{U_2\} \tag{3.128}$$

$$\{P\} = [K_{22}]\{U_2\} \tag{3.129}$$

$$\{R_2\} = [K_{32}]\{U_2\} \tag{3.130}$$

By inverting $[K_{22}]$, Equation (3.129) leads to

$$\{U_2\} = [K_{22}]^{-1}\{P\} \tag{3.131}$$

Then, from Equations (3.128) and (3.130), $\{R_1\}$ and $\{R_3\}$ become

$$\{R_1\} = [K_{12}][K_{22}]^{-1}\{P\} \tag{3.132}$$

and

$$\{R_3\} = [K_{32}][K_{22}]^{-1}\{P\} \tag{3.133}$$

By inspection of Equation (3.109), we see that $[K_{22}]$ is

$$[K_{22}] = 10^8 \begin{bmatrix} 12.39 & -2.277 & 28.19 \\ -2.277 & 3.209 & -8.782 \\ 28.19 & -8.782 & 93.15 \end{bmatrix} \tag{3.134}$$

Then $[K_{22}]^{-1}$ is found to be

$$[K_{22}]^{-1} = 10^{-9} \begin{bmatrix} 2.629 & -0.4204 & 0.8353 \\ -0.4204 & 4.268 & 0.5296 \\ -0.8353 & 0.5296 & 0.4101 \end{bmatrix} \tag{3.135}$$

Hence, from Equation (3.131), the unknown displacements are

$$U_{2x}/P = 2.629 \times 10^{-9} \text{ m/N}$$

$$U_{2y}/P = -0.4204 \times 10^{-9} \text{ m/N} \tag{3.136}$$

$$\theta_2/P = -0.8353 \times 10^{-9} \text{ rad/N}$$

From Equations (3.132) and (3.133) the unknown reactions are

$$R_{1x}/P = -7.0 \times 10^{-2}$$

$$R_{1y}/P = -2.18 \times 10^{-2}$$

$$M_1/P = 9.49 \times 10^{-1} \text{ m}$$

$$R_{3x}/P = -9.3 \times 10^{-1}$$

$$R_{3y}/P = 2.18 \times 10^{-2}$$

$$M_3/P = 2.942 \text{ m}$$

(3.137)

Now, to illustrate the ideas associated with joint releases, consider a beam-frame system similar to the system of Figure 3.31, but with the modification of having a pin at joint 2 as shown in Figure 3.33. There are three approaches we can take with this example. First, we can consider the pin as rigidly part of element (1) and hinged to element (2) as shown in Figure 3.34. Second, we can consider the pin as rigidly part of element (2) and hinged to element (1) as shown in Figure 3.35. Finally, we can consider the pin as hinged to both elements as originally shown in Figure 3.33.

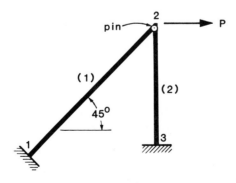

Figure 3.33 Simple Frame with a Pin at Joint 2

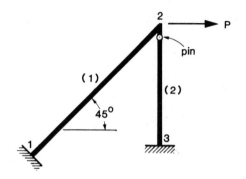

Figure 3.34 Simple Frame with Pin Attached to
Element (1) and Hinged to Element (2)

In the first case, that is, the case of Figure 3.34, the stiffness
matrix of element (1) referred to the global directions is the same as
in the preceding example. That is,

$$
\overset{(1)}{[k]} = 10^8
\begin{bmatrix}
2.691 & -2.277 & -8.782 & -2.691 & 2.277 & -8.782 \\
-2.277 & 2.691 & 8.782 & 2.277 & -2.691 & 8.782 \\
-8.782 & 8.782 & 41.40 & 8.782 & -8.782 & 20.70 \\
-2.691 & 2.277 & 8.782 & 2.691 & -2.277 & 8.782 \\
2.277 & -2.691 & -8.782 & -2.277 & 2.691 & -8.782 \\
-8.782 & 8.782 & 20.70 & 8.782 & -8.782 & 41.40
\end{bmatrix}
$$

$$(3.138)$$

where, as before, end 1 is at joint 1 and end 2 is at joint 2. The
stiffness matrix of element (2), however, is different than that of
the preceding example because of the hinge. If we consider joint 3
to be at end 1 of the element and joint 2 at end 2, then the local
stiffness matrix may be obtained directly from Equation (3.92).
That is,

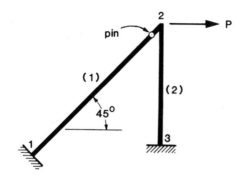

Figure 3.35 Simple Frame with Pin Attached to Element (2) and Hinged to Element (1)

$$
\overset{(2)}{[k]} = 10^8
\begin{bmatrix}
0.518 & 0.0 & 0.0 & -0.518 & 0.0 & 0.0 \\
0.0 & 2.426 & 9.703 & 0.0 & -2.426 & 0.0 \\
0.0 & 9.703 & 38.81 & 0.0 & -9.703 & 0.0 \\
-0.518 & 0.0 & 0.0 & 0.518 & 0.0 & 0.0 \\
0.0 & -2.426 & -9.703 & 0.0 & 2.426 & 0.0 \\
0.0 & 0.0 & 0.0 & 0.0 & 0.0 & 0.0
\end{bmatrix}
$$

$$(3.139)$$

Transformed relative to the global directions, this becomes

$$
\overset{(2)}{[k]} = 10^8
\begin{bmatrix}
2.426 & 0.0 & -9.703 & -2.426 & 0.0 & 0.0 \\
0.0 & 0.518 & 0.0 & 0.0 & -0.518 & 0.0 \\
-9.703 & 0.0 & 38.81 & 9.703 & 0.0 & 0.0 \\
-2.426 & 0.0 & 9.703 & 2.426 & 0.0 & 0.0 \\
0.0 & -0.518 & 0.0 & 0.0 & 0.518 & 0.0 \\
0.0 & 0.0 & 0.0 & 0.0 & 0.0 & 0.0
\end{bmatrix}
$$

$$(3.140)$$

Hence, by partitioning and superposition, as in the preceding example, the global stiffness matrix for the system is

$$[K] = 10^8 \begin{bmatrix} 2.691 & -2.277 & -8.782 & -2.691 & 2.277 & -8.782 & 0.0 & 0.0 & 0.0 \\ -2.277 & 2.691 & 8.782 & 2.277 & -2.691 & 8.782 & 0.0 & 0.0 & 0.0 \\ -8.782 & 8.782 & 41.40 & 8.782 & -8.782 & 20.70 & 0.0 & 0.0 & 0.0 \\ -2.691 & 2.277 & 8.782 & 5.117 & -2.277 & 8.782 & -2.426 & 0.0 & 9.703 \\ 2.277 & -2.691 & -8.782 & -2.277 & 3.209 & -8.782 & 0.0 & -0.5175 & 0.0 \\ -8.782 & 8.782 & 20.70 & 8.782 & -8.782 & 41.40 & 0.0 & 0.0 & 0.0 \\ 0.0 & 0.0 & 0.0 & -2.426 & 0.0 & 0.0 & 2.426 & 0.0 & -9.703 \\ 0.0 & 0.0 & 0.0 & 0.0 & -0.5175 & 0.0 & 0.0 & 0.5175 & 0.0 \\ 0.0 & 0.0 & 0.0 & 9.703 & 0.0 & 0.0 & -9.703 & 0.0 & 38.81 \end{bmatrix}$$

$$(3.141)$$

The force-displacement relation is the same as in the preceding example; that is, $\{F\}$ and $\{U\}$ are given by Equation (3.126). It should be noted, however, that in this case θ refers to the rotation of end 2 of element (1) and not the rotation of end 2 of element (2). That is, since element (2) is pinned at joint 2, with the corresponding joint release, its rotation at joint 2 is undetermined (see Problem 3.24). The unknown force and displacement components are then obtained directly by following the procedure outlined by Equations (3.126 3.133). In this case $[K_{22}]$ is found from Equation (3.141) to be

$$[K_{22}] = 10^8 \begin{bmatrix} 5.117 & -2.277 & 8.782 \\ -2.277 & 3.209 & -8.782 \\ 8.782 & -8.782 & 41.40 \end{bmatrix} \qquad (3.143)$$

Then $[K_{22}]^{-1}$ is found to be

$$[K_{22}]^{-1} = 10^{-9} \begin{bmatrix} 3.1984 & 0.9841 & -0.470 \\ 0.9841 & 7.735 & 1.432 \\ -0.470 & 1.432 & 0.645 \end{bmatrix} \qquad (3.144)$$

Hence, from Equation (3.131), the unknown displacements are found to be

$$U_{2x}/P = 3.20 \times 10^{-9} \text{ m/N}$$

$$U_{2y}/P = 0.984 \times 10^{-9} \text{ m/N} \qquad (3.145)$$

$$\theta_2/P = -0.470 \times 10^{-9} \text{ rad/N} \quad [\text{element (1)}]$$

From Equations (3.132) and (3.133) the unknown reactions are found to be

$$R_{1x}/P = -2.24 \times 10^{-1}$$

$$R_{1y}/P = 5.094 \times 10^{-2}$$

$$M_1/P = 0.972 \text{ m}$$

$$\qquad (3.146)$$

$$R_{3x}/P = -0.776$$

$$R_{3y}/P = -5.090 \times 10^{-2}$$

$$M_3/P = 3.104 \text{ m}$$

The second case, where the pin is attached to element (2), as shown in Figure 3.35, is similar to the first case. Indeed, the procedure is identical and therefore this case is left as an exercise (see Problem 3.24).

The third case, where the pin is assumed to be hinged to both elements (1) and (2), may be analyzed by using Equation (3.93) to obtain the local stiffness matrices of both elements. Specifically, the local and rotated stiffness matrix for element (2) are the same as those given in Equations (3.139) and (3.140). For element (1), if we consider joint 1 to be at end 1 and joint 2 to be at end 2, we find the local stiffness matrix to be

$$
\overset{(1)}{[k]} = 10^8
\begin{bmatrix}
0.414 & 0.0 & 0.0 & -0.414 & 0.0 & 0.0 \\
0.0 & 1.242 & 6.21 & 0.0 & -1.242 & 0.0 \\
0.0 & 6.21 & 31.05 & 0.0 & -6.21 & 0.0 \\
-0.414 & 0.0 & 0.0 & 0.414 & 0.0 & 0.0 \\
0.0 & -1.242 & -6.21 & 0.0 & 1.242 & 0.0 \\
0.0 & 0.0 & 0.0 & 0.0 & 0.0 & 0.0
\end{bmatrix}
$$

$$(3.147)$$

Rotated to the global directions, this becomes

$$
\overset{(1)}{[k]} = 10^8
\begin{bmatrix}
0.828 & -0.414 & -4.39 & -0.828 & 0.414 & 0.0 \\
-0.414 & 0.828 & 4.39 & 0.414 & -0.828 & 0.0 \\
-4.39 & 4.39 & 31.05 & 4.39 & -4.39 & 0.0 \\
-0.828 & 0.414 & 4.39 & 0.828 & -0.414 & 0.0 \\
0.414 & -0.828 & -4.39 & -0.414 & 0.828 & 0.0 \\
0.0 & 0.0 & 0.0 & 0.0 & 0.0 & 0.0
\end{bmatrix}
$$

$$(3.148)$$

The global stiffness matrix for the system is then

$$
[K] = 10^8
\begin{bmatrix}
0.828 & -0.414 & -4.391 & -0.828 & 0.414 & 0.0 & 0.0 & 0.0 & 0.0 \\
-0.414 & 0.828 & 4.391 & 0.414 & -0.828 & 0.0 & 0.0 & 0.0 & 0.0 \\
-4.391 & 4.391 & 31.05 & 4.391 & -4.391 & 0.0 & 0.0 & 0.0 & 0.0 \\
-0.828 & 0.414 & 4.391 & 3.254 & -0.414 & 0.0 & -2.426 & 0.0 & 9.703 \\
0.414 & -0.828 & -4.39 & -0.414 & 1.346 & 0.0 & 0.0 & -0.5175 & 0.0 \\
0.0 & 0.0 & 0.0 & 0.0 & 0.0 & 0.0 & 0.0 & 0.0 & 0.0 \\
0.0 & 0.0 & 0.0 & -2.426 & 0.0 & 0.0 & 2.426 & 0.0 & -9.703 \\
0.0 & 0.0 & 0.0 & 0.0 & -0.5175 & 0.0 & 0.0 & 0.5175 & 0.0 \\
0.0 & 0.0 & 0.0 & 9.703 & 0.0 & 0.0 & -9.703 & 0.0 & 38.81
\end{bmatrix}
$$

$$(3.149)$$

If we follow the procedures of the preceding examples in setting up the force-displacement relations as in Equation (3.125), and if we partition the matrices as in Equation (3.127), we find that $[K_{22}]$ is

$$[K_{22}] = 10^8 \begin{bmatrix} 3.254 & -0.414 & 0.0 \\ -0.414 & 1.346 & 0.0 \\ 0.0 & 0.0 & 0.0 \end{bmatrix} \qquad (3.150)$$

But, by inspection $[K_{22}]$ is seen to be singular and thus $[K_{22}]^{-1}$ is undefined. Therefore, we cannot use Equation (3.131) (that is, $\{U_2\} = [K_{22}]^{-1}\{P\}$) to determine $\{U_2\}$. However, if we reexamine Equation (3.129), we can still solve for U_{2x} and U_{2y} as follows: Let Equation (3.129) be rewritten in the partitioned form

$$\begin{bmatrix} P \\ 0 \\ \hline 0 \end{bmatrix} = \left[\begin{array}{cc|c} 3.254 & -0.414 & 0.0 \\ -0.414 & 1.346 & 0.0 \\ \hline 0.0 & 0.0 & 0.0 \end{array} \right] \begin{bmatrix} U_{2x} \\ U_{2y} \\ \hline \theta_2 \end{bmatrix} \qquad (3.151)$$

Hence, by block multiplication, we obtain

$$\begin{bmatrix} P \\ 0 \end{bmatrix} = 10^8 \begin{bmatrix} 3.254 & -0.414 \\ -0.414 & 1.346 \end{bmatrix} \begin{bmatrix} U_{2x} \\ U_{2y} \end{bmatrix} \qquad (3.152)$$

Then, by inverting the 2 × 2 matrix, we obtain

$$\begin{bmatrix} U_{2x} \\ U_{2y} \end{bmatrix} = 10^{-9} \begin{bmatrix} 3.20 & 0.984 \\ 0.984 & 7.735 \end{bmatrix} \begin{bmatrix} P \\ 0 \end{bmatrix} \qquad (3.153)$$

Or, finally

$$U_{2x}/P = 3.20 \times 10^{-9} \text{ M/N}$$
$$U_{2y}/P = 0.984 \times 10^{-9} \text{ M/N} \qquad (3.154)$$

The unknown reactions at joints 1 and 3 may now be determined by using Equations (3.128) and (3.130) as follows: Let Equation (3.128) be rewritten in partitioned form

$$\begin{bmatrix} R_{1x} \\ R_{1y} \\ M_1 \end{bmatrix} = 10^8 \left[\begin{array}{cc|c} -0.828 & 0.414 & 0.0 \\ 0.414 & -0.828 & 0.0 \\ \hline 4.391 & -4.391 & 0.0 \end{array}\right] \begin{bmatrix} U_{2x} \\ U_{2y} \\ H_2 \end{bmatrix} \quad (3.155)$$

where the numerical values for the matrix $[K_{12}]$ were obtained by inspection of Equation (3.149) in comparison with Equation (3.127). Then, by block matrix multiplication and using Equation (3.154), R_{1x}, R_{1y}, and M_1 are found to be

$$R_{1x}/P = -2.24 \times 10^{-1}$$
$$R_{1y}/P = 5.094 \times 10^{-2} \quad (3.156)$$
$$M_1/P = 0.972 \text{ m}$$

Similarly, by rewriting Equation (3.130) in partitioned form and using Equation (3.154), R_{3x}, R_{3y}, and M_3 are found to be

$$R_{3x}/P = -0.776$$
$$R_{3y}/P = -5.090 \times 10^{-2} \quad (3.157)$$
$$M_3/P = 3.104 \text{ m}$$

The results of Equations (3.154), (3.156), and (3.157) are, of course, consistent with the results given in Equations (3.145) and (3.146). The difference, however, is that θ_2 is not determined in this case. That is, θ_2 is undefined. But, this is consistent with the notion of leaving the pin hinged to *both* elements (1) and (2), because then the pin is unrestrained in rotation. Moreover, the frame in this case, is unable to sustain a moment at joint 2.

Finally, it might be observed that the computations in this third case were a bit simpler than in the first two cases. This is due to the additional zero's in the stiffness matrix of Equation (3.149). This advantage, however, is offset by the above-mentioned disadvantage of not being able to determine θ_2.

3.9 EXAMPLE: CHAPTER ILLUSTRATION FRAME

We are now finally ready to consider the example frame of Figure 3.1 which we have mentioned throughout the chapter. The frame is shown in Figure 3.36. All segments of the frame have identical cross sections with area $A = 10^{-3} m^2$, second moment of area $I = 0.025 m^4$, and all members have the same elastic properties with modulus $E = 2.07 \times 10^{11} N/m^2$.

In view of the foregoing discussion and example problems, the analysis of this frame is now essentially routine and straightforward.

Let the frame be divided into elements with the joints and elements numbered as shown in Figure 3.37. Let us also assume that the pin at joint 2 is rigidly attached to elements (1) and (3) and, therefore, hinged to element (2). The pin and roller at joint 4 simply mean that the reaction moment at the joint and the force along the roller support, are zero. We will assume that the pin at the roller is rigidly attached to element (3) and that the reaction moment at the roller is zero. [Alternatively, we could assume that the pin at the roller is *hinged* to element 3—thus simulating a *moment release* at that end of the element. The results with either

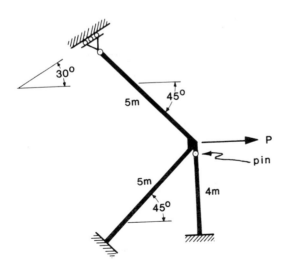

Figure 3.36 Example Frame

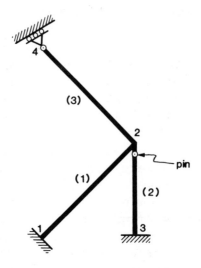

Figure 3.37   Joint and Element Numbering
for the Example Frame

assumption should, of course, be identical. Therefore, the procedure with this second assumption is left as an exercise (see Problem 3.30)].

To begin the analysis, assume that the "1" and "2" ends of the elements are associated with the joints as shown in Table 3.1. Then, in view of the above assumptions on the pin at joint 2, Table 3.1 shows that element (2) has a moment release at end 2. Hence, by using Equations (3.34) and (3.93) the local element stiffness matrices are found to be

$$
\overset{(1)}{[k]} = 10^8
\begin{bmatrix}
0.414 & 0.0 & 0.0 & -0.414 & 0.0 & 0.0 \\
0.0 & 4.968 & 12.42 & 0.0 & -4.968 & 12.42 \\
0.0 & 12.42 & 41.4 & 0.0 & -12.42 & 20.70 \\
-0.417 & 0.0 & 0.0 & 0.414 & 0.0 & 0.0 \\
0.0 & -4.968 & -12.42 & 0.0 & 4.968 & -12.42 \\
0.0 & 12.42 & 20.70 & 0.0 & -12.42 & 41.40
\end{bmatrix}
$$

$$(3.158)$$

Table 3.1 Identification of Element Ends and Joint Numbers

Element No.	Joint Number	
	End 1	End 2
(1)	1	2
(2)	3	2
(3)	4	2

$$
\overset{(2)}{[k]} = 10^8
\begin{bmatrix}
0.5175 & 0.0 & 0.0 & -0.5175 & 0.0 & 0.0 \\
0.0 & 2.426 & 9.703 & 0.0 & -2.426 & 0.0 \\
0.0 & 9.703 & 38.81 & 0.0 & -9.703 & 0.0 \\
-0.5175 & 0.0 & 0.0 & 0.5175 & 0.0 & 0.0 \\
0.0 & -2.426 & -9.703 & 0.0 & 2.426 & 0.0 \\
0.0 & 0.0 & 0.0 & 0.0 & 0.0 & 0.0
\end{bmatrix}
$$

(3.159)

$$
\overset{(3)}{[k]} = 10^8
\begin{bmatrix}
0.414 & 0.0 & 0.0 & -0.414 & 0.0 & 0.0 \\
0.0 & 4.968 & 12.42 & 0.0 & -4.968 & 12.42 \\
0.0 & 12.42 & 41.40 & 0.0 & -12.42 & 20.70 \\
-0.414 & 0.0 & 0.0 & 0.414 & 0.0 & 0.0 \\
0.0 & -4.968 & -12.42 & 0.0 & 4.968 & -12.42 \\
0.0 & 12.42 & 20.70 & 0.0 & -12.42 & 41.40
\end{bmatrix}
$$

(3.160)

where, as in the foregoing examples, the units are N/m, N, and Nm depending upon the row and column.

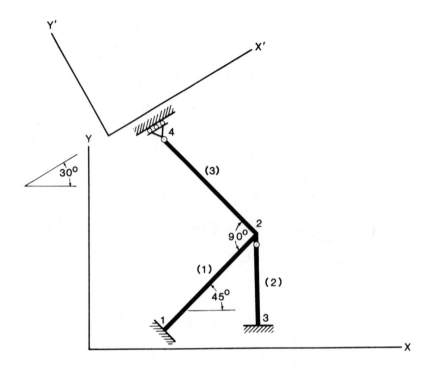

Figure 3.38 Global Directions for the Example Frame

Let the global directions be the horizontal and vertical x and y axes as shown in Figure 3.38. However, at joint 4, let the global directions be along and perpendicular to the roller track, that is, along X' and Y' as also shown in Figure 3.38. Hence, for element (1) the inclination angle γ is $45°$. For element (2), γ is $90°$. For element (3), γ is $-75°$ at joint 4 (end 1) and $-45°$ at joint 2 (end 2). Then by using Equation (3.50), and the results of Problem 3.8, the local stiffness matrices referred to the global directions become

$$
\overset{(1)}{[k]} = 10^8
\begin{bmatrix}
2.691 & -2.277 & -8.782 & -2.691 & 2.277 & -8.782 \\
-2.277 & 2.691 & 8.782 & 2.277 & -2.691 & 8.782 \\
-8.782 & 8.782 & 41.40 & 8.782 & -8.782 & 20.70 \\
-2.691 & 2.277 & 8.782 & 2.691 & -2.277 & 8.782 \\
2.277 & -2.691 & -8.782 & -2.277 & 2.691 & -8.782 \\
-8.782 & 8.782 & 20.70 & 8.782 & -8.782 & 41.40
\end{bmatrix}
$$

$$(3.161)$$

$$
\overset{(2)}{[k]} = 10^8
\begin{bmatrix}
2.426 & 0.0 & -9.703 & -2.426 & 0.0 & 0.0 \\
0.0 & 0.5175 & 0.0 & 0.0 & -0.517 & 0.0 \\
-9.703 & 0.0 & 38.81 & 9.703 & 0.0 & 0.0 \\
-2.426 & 0.0 & 9.703 & 2.426 & 0.0 & 0.0 \\
0.0 & -0.5175 & 0.0 & 0.0 & 0.517 & 0.0 \\
0.0 & 0.0 & 0.0 & 0.0 & 0.0 & 0.0
\end{bmatrix}
$$

(3.162)

and

$$
\overset{(3)}{[k]} = 10^8
\begin{bmatrix}
4.663 & 1.139 & 12.00 & -3.469 & -3.317 & 12.00 \\
1.139 & 0.719 & 3.215 & -0.624 & -1.192 & 3.215 \\
12.00 & 3.215 & 41.40 & -8.782 & -8.782 & 20.70 \\
-3.469 & -0.626 & -8.782 & 2.691 & 2.277 & -8.782 \\
-3.317 & -1.192 & -8.782 & 2.277 & 2.691 & -8.782 \\
12.00 & 3.215 & 20.70 & -8.782 & -8.782 & 41.40
\end{bmatrix}
$$

(3.163)

We are now ready to assemble the system by superposing the element stiffness matrices. Specifically, by using the procedure of Section 3.5, by referring to Table 3.1, by expanding the rotated element stiffness matrices of Equations (3.161), (3.162), and (3.163) into global dimensions, and finally, by adding, the global stiffness matrix $[K]$ is obtained.

To facilitate this it is convenient to follow the same procedure adopted in Section 3.8 where we partitioned the local element stiffness matrices into submatrices. Hence, in Equations (3.161) to (3.163), let the stiffness matrices be written in the form

$$
\overset{(e)}{[k]} =
\left[
\begin{array}{c|c}
\overset{(e)}{K_{11}} & \overset{(e)}{K_{12}} \\
\hline
\overset{(e)}{K_{21}} & \overset{(e)}{K_{22}}
\end{array}
\right]
\quad (e = 1, 2, 3)
$$

(3.164)

Then the $[\kappa_{ij}]^{(e)}$ are identified by inspection and comparison of Equations (3.161) to (3.163) and Equation (3.164). The global stiffness matrix $[K]$ is a 12×12 array. It may be partitioned into 16 3×3 submatrices as follows:

$$[K] = \begin{bmatrix} K_{11} & K_{12} & K_{13} & K_{14} \\ K_{21} & K_{22} & K_{23} & K_{24} \\ K_{31} & K_{32} & K_{33} & K_{34} \\ K_{41} & K_{42} & K_{43} & K_{44} \end{bmatrix} \tag{3.165}$$

Then, by using the outlined procedures, the $[K_{ij}]$, $i, j = 1, \ldots, 4$ are found to be

$$[K_{11}] = [\kappa_{11}^{(1)}] = 10^8 \begin{bmatrix} 2.691 & -2.277 & -8.782 \\ -2.277 & 2.691 & 8.782 \\ -8.782 & 8.782 & 41.40 \end{bmatrix} \tag{3.166}$$

$$[K_{12}] = [K_{21}]^T = [\kappa_{12}^{(1)}] = 10^8 \begin{bmatrix} -2.691 & 2.277 & -8.782 \\ 2.277 & -2.691 & 8.782 \\ 8.782 & -8.782 & 20.70 \end{bmatrix} \tag{3.167}$$

$$[K_{13}] = [K_{31}]^T = 10^8 \begin{bmatrix} 0.0 & 0.0 & 0.0 \\ 0.0 & 0.0 & 0.0 \\ 0.0 & 0.0 & 0.0 \end{bmatrix} \tag{3.168}$$

$$[K_{14}] = [K_{41}]^T = 10^8 \begin{bmatrix} 0.0 & 0.0 & 0.0 \\ 0.0 & 0.0 & 0.0 \\ 0.0 & 0.0 & 0.0 \end{bmatrix} \tag{3.169}$$

$$[K_{22}] = [\overset{(1)}{\kappa_{22}}] + [\overset{(2)}{\kappa_{22}}] + [\overset{(3)}{\kappa_{22}}] = 10^8 \begin{bmatrix} 7.808 & 0.0 & 0.0 \\ 0.0 & 5.90 & -17.56 \\ 0.0 & -17.56 & 82.80 \end{bmatrix}$$

$$(3.170)$$

$$[K_{23}] = [K_{32}]^T = [\overset{(2)}{\kappa_{21}}] = 10^8 \begin{bmatrix} -2.426 & 0.0 & 9.703 \\ 0.0 & -0.518 & 0.0 \\ 0.0 & 0.0 & 0.0 \end{bmatrix} \quad (3.171)$$

$$[K_{24}] = [K_{42}]^T [\overset{(3)}{\kappa_{21}}] = 10^8 \begin{bmatrix} -3.469 & -0.626 & -8.782 \\ -3.317 & -1.192 & -8.782 \\ 12.00 & 3.215 & 20.70 \end{bmatrix} \quad (3.172)$$

$$[K_{33}] = [\overset{(2)}{\kappa_{11}}] = 10^8 \begin{bmatrix} 2.426 & 0.0 & -9.703 \\ 0.0 & 0.5175 & 0.0 \\ -9.703 & 0.0 & 38.81 \end{bmatrix} \quad (3.173)$$

$$[K_{34}] = [K_{43}]^T = 10^8 \begin{bmatrix} 0.0 & 0.0 & 0.0 \\ 0.0 & 0.0 & 0.0 \\ 0.0 & 0.0 & 0.0 \end{bmatrix} \quad (3.174)$$

$$[K_{44}] = [\overset{(3)}{\kappa_{11}}] = 10^8 \begin{bmatrix} 4.663 & 1.139 & 12.00 \\ 1.139 & 0.719 & 3.215 \\ 12.00 & 3.215 & 41.40 \end{bmatrix} \quad (3.175)$$

The force-displacement equation is of the usual form, that is,

$$\{F\} = [K]\{U\} \qquad (3.176)$$

where in this case it is convenient to write $\{F\}$ and $\{U\}$ as

$$\{F\} = \begin{bmatrix} R_1 \\ \hline P \\ \hline R_3 \\ \hline R_R \end{bmatrix} = \begin{bmatrix} R_{1x} \\ R_{1y} \\ M_1 \\ \hline R_{2x} \\ R_{2y} \\ M_2 \\ \hline R_{3x} \\ R_{3y} \\ M_3 \\ \hline R_{4x}' \\ R_{4y}' \\ M_4 \end{bmatrix} = \begin{bmatrix} R_{1x} \\ R_{1y} \\ M_1 \\ \hline P \\ 0 \\ 0 \\ \hline R_{3x} \\ R_{3y} \\ M_3 \\ \hline 0 \\ R \\ 0 \end{bmatrix} \qquad (3.177)$$

and

$$\{U\} = \begin{bmatrix} U_1 \\ \hline U_2 \\ \hline U_3 \\ \hline U_R \end{bmatrix} = \begin{bmatrix} U_{1x} \\ U_{1y} \\ \theta_1 \\ \hline U_{2x} \\ U_{2y} \\ \theta_2 \\ \hline U_{3x} \\ U_{3y} \\ \theta_3 \\ \hline U_{4x}' \\ U_{4y}' \\ \theta_4 \end{bmatrix} = \begin{bmatrix} 0 \\ 0 \\ 0 \\ \hline U_{2x} \\ U_{2y} \\ \theta_2 \\ \hline 0 \\ 0 \\ 0 \\ \hline U_R \\ 0 \\ \theta_4 \end{bmatrix} \qquad (3.178)$$

where the notation is meant to be self-suggestive. That is, R is the reaction force at joint 4, normal to the roller track, and U_R is the displacement along the roller track.

Substituting these expressions into Equation (3.176), it is seen that the matrix equation is equivalent to 12 scalar equations. Thus, the system is well defined since there are 12 unknowns as seen in Equations (3.177), and (3.178): R_{1x}, R_{1y}, M_1, P, R_{3x}, R_{3y}, M_3, R, U_{2x}, U_{2y}, θ_2, U_R, and θ_4.

The system may be solved as follows: By block multiplication of the terms of Equations (3.165), (3.177), and (3.178) and by collecting the scalar terms representing the known force components, the following expression is obtained:

$$
\begin{bmatrix} P \\ 0 \\ 0 \\ 0 \\ 0 \end{bmatrix} = \begin{bmatrix} K_{44} & K_{45} & K_{46} & K_{4,10} & K_{4,12} \\ K_{54} & K_{55} & K_{56} & K_{5,10} & K_{5,12} \\ K_{64} & K_{65} & K_{66} & K_{6,10} & K_{6,12} \\ K_{10,4} & K_{10,5} & K_{10,6} & K_{10,10} & K_{10,12} \\ K_{12,4} & K_{12,5} & K_{12,6} & K_{12,10} & K_{12,12} \end{bmatrix} \begin{bmatrix} U_{2x} \\ U_{2y} \\ \theta_2 \\ U_R \\ \theta_4 \end{bmatrix} \quad (3.179)
$$

where the K_{ij} ($i, j = 1, \ldots, 12$) are elements of $[K]$. This expression is conveniently written in the compact form

$$\{\hat{P}\} = [\hat{K}] \{\hat{U}\} \quad (3.180)$$

where $\{\hat{P}\}$, $[\hat{K}]$, and $\{\hat{U}\}$ are defined by comparison of Equations (3.179) and (3.180).

By solving Equation (3.180) for $\{\hat{U}\}$ we obtain

$$\{\hat{U}\} = [\hat{K}]^{-1} \{\hat{P}\} \quad (3.181)$$

From Equations (3.165) to (3.175) $[\hat{K}]$ is seen to be

$$
[\hat{K}] = 10^8 \begin{bmatrix} 7.808 & 0.0 & 0.0 & -3.469 & -8.782 \\ 0.0 & 5.900 & -17.56 & -3.317 & -8.782 \\ 0.0 & -17.56 & 82.80 & 12.00 & 20.70 \\ -3.469 & -3.317 & 12.00 & 4.663 & 12.00 \\ -8.782 & -8.782 & 20.70 & 12.00 & 41.40 \end{bmatrix}
$$

$$(3.182)$$

Then $[\hat{K}]^{-1}$ is

$$[\hat{K}]^{-1} = 10^{-9} \begin{bmatrix} 3.189 & 1.005 & -0.465 & 5.490 & -0.468 \\ 1.005 & 6.989 & 1.322 & -1.348 & 1.427 \\ -0.465 & 1.322 & 0.632 & -2.674 & 0.642 \\ 5.490 & -1.348 & -2.674 & 25.32 & -5.123 \\ -0.468 & 1.427 & 0.642 & -5.123 & 1.604 \end{bmatrix}$$

$$(3.183)$$

The unknown displacements are then obtained from Equation (3.181) as

$$U_{2x}/P = 3.189 \times 10^{-9} \text{ m/N}$$

$$U_{2y}/P = 1.005 \times 10^{-9} \text{ m/N}$$

$$\theta_2/P = -4.65 \times 10^{-10} \text{ rad/N}$$ (3.184)

$$U_R/P = 5.490 \times 10^{-9} \text{ m/N}$$

$$\theta_4/P = -4.68 \times 10^{-10} \text{ rad/N}$$

By using Equation (3.176), together with Equations (3.166) to (3.175) and (3.184), the unknown forces are determined as

$$R_{1x}/P = -0.2209$$

$$R_{1y}/P = 0.04917$$

$$M_1/P = 0.958 \text{ m}$$

$$R_{3x}/P = -0.739$$ (3.185)

$$R_{3y}/P = -0.0517$$

$$M_3/P = 3.094 \text{ m}$$

$$R/P = 5.66 \times 10^{-3}$$

3.10   GENERALIZATION TO THREE DIMENSIONS

In this section we will briefly consider generalizing the foregoing ideas to three-dimensional beam and frame elements. Conceptually, the ideas are the same. The only change is the introduction of additional terms due to the third dimension. These terms include the effect of torsion along the element axis. The specific procedures for determining the element stiffness matrix, for assembly, for determining fixed-end loads, and for joint releases, are identical.

To begin, consider the beam element as shown in Figure 3.39. Let there be force systems applied at each end of the element. Let these force systems be equivalent to single forces f_i ($i = 1, 2$) passing through the intersection points of the x axis and the end cross-sections, together with couples with moments m_i ($i = 1, 2$), where the subscripts $(1, 2)$ refer to the respective ends of the beam element. For example, at end 1, let f_1 and m_1 be written

$$f_1 = f_{1x}\, n_x + f_{1y} n_y + f_{1z}\, n_z \tag{3.186}$$

and

$$m_1 = m_{1x}\, n_x + m_{1y}\, n_y + m_{1z}\, n_z \tag{3.187}$$

where the unit vectors n_x, n_y, and n_z are parallel to the X, Y, and Z axes as shown in Figure 3.39. The scalar components of f_1 and m_1 in Equations (3.186) and (3.187) are, of course, the force and moment components along the X, Y, and Z axes. Hence, m_{1x} is the twisting or torsional moment on the beam element.

Analogously, if we assume that the displacements resulting from these force systems are small, we can let the displacement be represented by a translation u_i ($i = 1, 2$) together with a rotation θ_i ($i = 1, 2$), where u_i and θ_i are

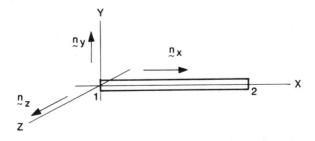

Figure 3.39   Three-Dimensional Beam Element

$$u_i = u_{1x}\, n_x + u_{1y}\, n_y + u_{1z}\, n_z \qquad (3.188)$$

and

$$\boldsymbol{\theta}_i = \theta_{1x}\, n_x + \theta_{1y}\, n_y + \theta_{1z}\, n_z \qquad (3.189)$$

In Equation (3.189), θ_{1x}, θ_{1y}, and θ_{1z} are the rotations at end 1 about the X, Y, and Z axes respectively, with θ_{1x} being the twist due to the torsion moment m_{1x} exerted on the element.

By assuming that the translation and rotations are small, we can use the principle of superposition, as in Section 3.3 to obtain the relationships between the force and moment components and the translation and rotation components. That is, if the 12 force and moment components are listed in an array $\{f\}$, and if the 12 translation and rotation components are listed in an array $\{u\}$, then $\{f\}$ and $\{u\}$ are related by the familiar expression

$$\{f\} = [k]\,\{u\} \qquad (3.190)$$

where $[k]$ is now a 12×12 stiffness matrix. Specifically, if $\{f\}$ and $\{u\}$ are written

$$\{f\} = \begin{bmatrix} f_1 \\ f_2 \\ f_3 \\ f_4 \\ f_5 \\ f_6 \\ f_7 \\ f_8 \\ f_9 \\ f_{10} \\ f_{11} \\ f_{12} \end{bmatrix} = \begin{bmatrix} f_{1x} \\ f_{1y} \\ f_{1z} \\ m_{1x} \\ m_{1y} \\ m_{1z} \\ f_{2x} \\ f_{2y} \\ f_{2z} \\ m_{2x} \\ m_{2y} \\ m_{2z} \end{bmatrix} \quad \text{and} \quad \{u\} = \begin{bmatrix} u_1 \\ u_2 \\ u_3 \\ u_4 \\ u_5 \\ u_6 \\ u_7 \\ u_8 \\ u_9 \\ u_{10} \\ u_{11} \\ u_{12} \end{bmatrix} = \begin{bmatrix} u_{1x} \\ u_{1y} \\ u_{1z} \\ \theta_{1x} \\ \theta_{1y} \\ \theta_{1z} \\ u_{2x} \\ u_{2z} \\ u_{2z} \\ \theta_{2x} \\ \theta_{2y} \\ \theta_{2z} \end{bmatrix}$$

$$(3.191, 2)$$

$[k]$ is found to be (see Problem 3.31)

$$[k] = \begin{bmatrix}
AE/\ell & 0 & 0 & 0 & 0 & 0 & -AE/\ell & 0 & 0 & 0 & 0 & 0 \\
0 & 12EI_z/\ell^3 & 0 & 0 & 0 & 6EI_z/\ell^2 & 0 & -12EI_z/\ell^3 & 0 & 0 & 0 & 6EI_z/\ell^2 \\
0 & 0 & 12EI_y/\ell^3 & 0 & -6EI_y/\ell^2 & 0 & 0 & 0 & -12EI_y/\ell^3 & 0 & -6EI_y/\ell^2 & 0 \\
0 & 0 & 0 & GI_x/\ell & 0 & 0 & 0 & 0 & 0 & -GI_x/\ell & 0 & 0 \\
0 & 0 & -6EI_y/\ell^2 & 0 & 4EI_y/\ell & 0 & 0 & 0 & 6EI_y/\ell^2 & 0 & 2EI_y/\ell & 0 \\
0 & 6EI_z/\ell^2 & 0 & 0 & 0 & 4EI_z/\ell & 0 & -6EI_z/\ell^2 & 0 & 0 & 0 & 2EI_z/\ell \\
-AE/\ell & 0 & 0 & 0 & 0 & 0 & AE/\ell & 0 & 0 & 0 & 0 & 0 \\
0 & -12EI_z/\ell^3 & 0 & 0 & 0 & -6EI_z/\ell^2 & 0 & 12EI_z/\ell^3 & 0 & 0 & 0 & -6EI_z/\ell^2 \\
0 & 0 & -12EI_y/\ell^3 & 0 & 6EI_y/\ell^2 & 0 & 0 & 0 & 12EI_y/\ell^3 & 0 & 6EI_y/\ell^2 & 0 \\
0 & 0 & 0 & -GI_x/\ell & 0 & 0 & 0 & 0 & 0 & GI_x/\ell & 0 & 0 \\
0 & 0 & -6EI_y/\ell^2 & 0 & 2EI_y/\ell & 0 & 0 & 0 & 6EI_y/\ell^2 & 0 & 4EI_y/\ell & 0 \\
0 & 6EI_z/\ell^2 & 0 & 0 & 0 & 2EI_z/\ell & 0 & -6EI_z/\ell^2 & 0 & 0 & 0 & 4EI_z/\ell
\end{bmatrix} \tag{3.193}$$

where I_x, I_y, and I_z are second moments of area of the element cross section about the x, y, and z axes. (I_x is sometimes called the "second polar moment of area.") Finally, G is the shear modulus.

Equation (3.193) is directly analogous to Equation (3.34) of Section 3.3. Both of these equations are developed by assuming small displacement and by neglecting shear deformation.[1]

As mentioned above, the generalization of Equation (3.193) to oblique directions follows directly from the procedures of Section 3.4 and Problems 3.7 and 3.8 (see Problem 3.32). Similarly, the procedures of assembly, fixed-end loads, and joint releases are identical to those of Sections 3.5, 3.6, and 3.7.

It should be clear, however, from this brief discussion and from Equation (3.193) that although three-dimensional beam elements are conceptually similar to two-dimensional beam elements, they are computationally considerably more detailed. Hence, it is impractical to consider an analysis of a three-dimensional frame without the aid of a digital computer to perform the algebra of the arrays. It is beyond the scope of this introductory text to develop suitable computer algorithms to perform this analysis—although conceptually, such a development is routine and it follows directly from the procedures introduced and discussed herein. Fortunately, there already exist a number of commercially available computer programs[2] which automatically process the details of three-dimensional frame analyses. The reader is encouraged to explore the use and function of such programs as they are available at his or her institution or company.

PROBLEMS

Section 3.2

3.1 Verify Equations (3.1) and (3.2) from strength of materials theory.

---

[1] To see the effects of including shear deformation, see Przemieniecki, J. S., *Theory of Matrix Structural Analysis*, McGraw-Hill, New York, 1968, pp. 79-80.

[2] For example, STRUDL, SAP, and NASTRAN.

Section 3.3

3.2   Check the development of Equation (3.19) from Equation (3.18).

3.3   Derive Equations (3.21) from equilibrium considerations of the beam element of Figure 3.11.

3.4   Derive Equations (3.23) by adopting Equations (3.1), (3.2), and (3.5) for the beam element of Figure 3.11.

3.5   Show that the square matrix of Equation (3.22) is the inverse of the square matrix of Equation (3.16).

Section 3.4

3.6   Verify that $[T]$ defined in Equation (3.44) is orthogonal.

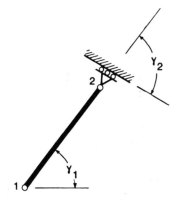

Figure P3.7   Beam Element with Inclined
Roller Support at End 2

3.7 Suppose a beam element is attached to an inclined roller at one of its ends, as, for example, in Figure P3.7. It may be desirable in such a case to transform the element quantities to *different* global directions at the respective ends of the element. Following the development in Section 3.4, show that in this case Equations (3.40) to (3.43) would become

$$\{u_1\} = [t_1]\{u_1'\}, \quad \{f_1\} = [t_1]\{f_1'\}$$

$$\{u_2\} = [t_2]\{u_2'\}, \quad \{f_2\} = [t_2]\{f_2'\}$$

where the subscripts on $[t]$ refer to the angles γ_1 and γ_2 of Figure P3.7. Further show that, analogous to Equations (3.45) and (3.49), $[T]$ and $[k]$ become

$$[T] = \begin{bmatrix} t_1 & | & 0 \\ --- & + & --- \\ 0 & | & t_2 \end{bmatrix}$$

and

$$[k] = [T][k'][T]^T = \begin{bmatrix} [t_1][\kappa'_{11}][t_1]^T & | & [t_1][\kappa'_{12}][t_2]^T \\ --- & + & --- \\ [t_2][\kappa'_{21}][t_1]^T & | & [t_2][\kappa'_{22}][t_2]^T \end{bmatrix}$$

where, as in Section 3.4 $[\kappa'_{ij}]$ are the partitioned blocks of k'.

3.8 Referring to Problem 3.7, show that the stiffness matrix $[k]$ analogous to Equation (3.50) for different global directions at the respective ends is as shown on the following page.

$$[k] = \frac{EI}{\ell^3}\begin{bmatrix}
\left(\dfrac{A\ell^2}{I}c_1^2+12s_1^2\right) & \left(\dfrac{A\ell^2}{I}s_1c_1-12s_1c_1\right) & (-6\ell s_1) & \left(-\dfrac{A\ell^2}{I}c_1c_2-12s_1s_2\right) & \left(-\dfrac{A\ell^2}{I}s_2c_1+12s_1c_2\right) & (-6\ell s_1) \\[2ex]
\left(\dfrac{A\ell^2}{I}s_1c_1-12s_1c_1\right) & \left(\dfrac{A\ell^2}{I}s_1^2+12c_1^2\right) & (6\ell c_1) & \left(-\dfrac{A\ell}{I}s_1c_1+12s_2c_1\right) & \left(-\dfrac{A\ell^2}{I}s_1s_2-12c_1c_2\right) & (6\ell c_1) \\[2ex]
(-6\ell s_1) & (6\ell c_1) & (4\ell^2) & (6\ell s_2) & (-6\ell c_2) & (2\ell^2) \\[2ex]
\left(-\dfrac{A\ell^2}{I}c_1c_2-12s_1s_2\right) & \left(-\dfrac{A\ell^2}{I}s_1c_2+12s_2c_1\right) & (6\ell s_2) & \left(\dfrac{A\ell^2}{I}c_2^2+12s_2^2\right) & \left(\dfrac{A\ell^2}{I}s_2c_2-12s_2c_2\right) & (6\ell s_2) \\[2ex]
\left(-\dfrac{A\ell^2}{I}s_2c_1+12s_1c_2\right) & \left(-\dfrac{A\ell^2}{I}s_1s_2-12c_1c_2\right) & (-6\ell c_2) & \left(\dfrac{A\ell^2}{I}s_2c_2-12s_2c_2\right) & \left(\dfrac{A\ell^2}{I}s_2^2+12c_2^2\right) & (-6\ell c_2) \\[2ex]
(-6\ell s_1) & (6\ell c_1) & (2\ell^2) & (6\ell s_2) & (-6\ell c_2) & (4\ell^2)
\end{bmatrix}$$

Section 3.5

3.9 Develop $[\overset{(1)}{\Lambda}]$ and $[\overset{(2)}{\Lambda}]$ of Equations (3.53) and (3.54) by writing a set of equations like Equations (3.57) and by developing "incidence tables" like Tables 2.1, 2.2, and 2.3, of Section 2.6.

3.10 Check the results of Equation (3.76).

3.11 Following the example of Section 3.5, determine the unknown displacements and reactions of a simply supported beam of length L with a concentrated load at a distance a from the left end (see Figure P3.11).

Section 3.6

3.12 Verify that the forces and moments of Equation (3.77) are indeed the fixed-end reactions of a uniformly loaded beam as in Figure 3.23. [*Hint*: Solve the differential equation

$$EI \frac{d^4 v}{dx^4} = -p_0$$

subject to the boundary conditions

$$v(0) = v(L) = \frac{dv}{dx}(0) = \frac{dv}{dx}(L) = 0.]$$

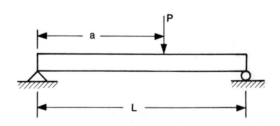

Figure P3.11 Simply Supported Beam with
a Concentrated Load

3.13 Using the procedure of Section 3.6 and following the example problem, find the displacements and rotations for a uniformly loaded fixed-end beam using two elements.

Section 3.7

3.14 Imagine the beam element of Figure 3.30 to be supported at end 1 and loaded at end 2. Show that

$$f_{1x} = -f_{2x}$$

$$f_{1y} = -f_{2y}$$

$$m_1 = -f_{2y}\ell = f_{1y}\ell$$

Then consider the elements to be pin supported at end 2 and loaded axially at end 1. Show that

$$u_{1x} = f_{1x}\ell/AE \quad \text{or} \quad f_{1x} = u_{1x}AE/\ell$$

3.15 Imagine the beam element of Figure 3.30 to be loaded at end 2 and supported at end 1. Show that u_{2x} and u_{2y} are related to f_{2x} and f_{2y} as

$$u_{2x} = f_{2x}\ell/AE \quad \text{and} \quad u_{2y} = f_{2y}\ell^3/3EI$$

or

$$f_{2x} = u_{2x}AE/\ell \quad \text{and} \quad f_{2y} = 3u_{2y}EI/\ell^3$$

and hence verify $[\hat{\kappa}_{22}]$ of Equation (3.91).

3.16 Combining the results of Problems 3.14 and 3.15 show that

$$f_{1x} = -u_{2x}AE/\ell$$

$$f_{1y} = -3u_{2y}EI/\ell^3$$

$$m_1 = -3u_{2y}EI/\ell^2$$

and hence verify $[\hat{\kappa}_{12}]$ of Equation (3.89). Using these results for $[\hat{\kappa}_{12}]$ and symmetry verify $[\hat{\kappa}_{21}]$ of Equation (3.90). That is, show that

$$f_{2x} = -u_{1x}AE/\ell$$

$$f_{2y} = -3u_{1y}EI/\ell^3 - 3\theta_1 EI/\ell^2$$

3.17   By using the results from Problem 3.16 [that is, Equation (3.90)] and the results of Problem 3.14, show that

$$f_{1x} = u_{1x}AE/\ell$$

$$f_{1y} = 3u_{1y}EI/\ell^3 + 3\theta_1 EI/\ell^2$$

$$m_1 = 3u_{1y}EI/\ell^2 + 3\theta_1 EI/\ell$$

Hence, verify $[\hat{\kappa}_{11}]$ of Equation (3.88).

3.18   Following the procedure outlined in Problems 3.14 through 3.17, show that if a beam element has a "moment release" at end 1, its 6×6 element stiffness matrix is

$$[k]_3 = \begin{bmatrix} AE/\ell & 0 & 0 & -AE/\ell & 0 & 0 \\ 0 & 3EI/\ell^3 & 0 & 0 & -3EI/\ell^3 & 3EI/\ell^2 \\ 0 & 0 & 0 & 0 & 0 & 0 \\ -AE/\ell & 0 & 0 & AE/\ell & 0 & 0 \\ 0 & -3EI/\ell^3 & 0 & 0 & 3EI/\ell^3 & -3EI/\ell^2 \\ 0 & 3EI/\ell^2 & 0 & 0 & -3EI/\ell^2 & 3EI/\ell \end{bmatrix}$$

3.19   Using Equations (3.95) to (3.99), show that the expression for $[\hat{k}]$ of Equation (3.106) is identical to the expression of Equation (3.92).

3.20   Obtain the 6×6 element stiffness matrix for a beam element with a moment release at end 1 (as in Problem 3.18), by following the procedure used to develop Equation (3.105).

Section 3.8

3.21   Verify the accuracy of the stiffness matrices of Equations (3.108) and (3.109).

3.22   Verify the terms of the global stiffness matrix of Equation (3.124) by superposing expanded forms of the stiffness matrices of Equations (3.108) and (3.109). (Note that for element (2), end 1 is at joint 3 and end 2 is at joint 2.)

3.23   Following the procedure outlined in the text, find the unknown displacements at joint 2 and the unknown reactions at joints 1 and 3 for the frame of Figure P3.23. Use the same geometrical and physical constants as in the frame of Figure 3.31.

3.24   Using the procedures outlined for Case 1 of the example of Figure 3.34, of Section 3.8, obtain the local stiffness matrices for Case 2 of Figure 3.35. Find the global stiffness matrix $[K]$ and determine the unknown displacements and reactions. Verify that the results are identical to those of Equations (1.45) and (1.46) except for θ which is now the end 2 rotation of element (2).

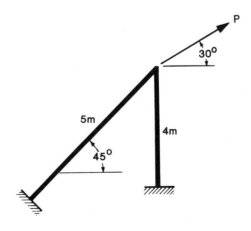

Figure P3.23 Simple Frame with Oblique Loading

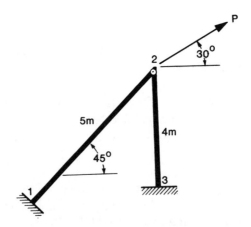

Figure P3.25   Example Frame with Oblique Loading

3.25 (See Problem 3.21) Following the procedures of the second example of Section 3.8, find the unknown displacement at joint 2 and the unknown reactions at joints 1 and 3 of the example of Figure 3.33 with the loading Figure P3.25.

3.26 (See Problem 3.22) Develop the global stiffness matrix of Equation (3.149) by superposing expanded forms of the stiffness matrices of Equations (3.140) and (3.148).

Section 3.9

3.27 Using Equation (3.50) and the results of Problem 3.8, verify the numerical values of Equations (3.161), (3.162), and (3.163) for the element stiffness matrices.

3.28 Using the element stiffness matrices of Equations (3.161), (3.162), and (3.163) verify the results of the assembly into the global stiffness matrix $[K]$ of Equations (3.165) to (3.175).

3.29 Using the procedures and results of Section 3.8, find the unknown displacement and reactions for the frame shown in Figure P3.29. (The physical data are the same as for the system of Figure 3.36.)

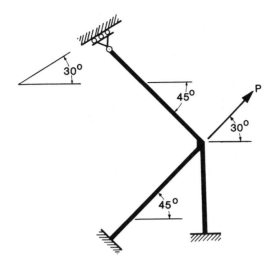

Figure P3.29 Example Frame with Oblique Loading

3.30 Rework the example problem of Section 3.9 by assuming that the pin at the roller is *hinged* to element (3), thus simulating a moment release at that end of the element. Compare the results and procedures to those of Section 3.9.

Section 3.10

3.31 Using the principle of superposition and the procedures of Section 3.3, verify the results of Equation (3.193).

3.32 Show that the three-dimensional beam element stiffness matrix $[k]$ of Equation (3.193), may be transformed to oblique coordinate directions for each end of the element as follows: Let n'_x, n'_y, and n'_z be mutually perpendicular unit vectors along the convenient oblique directions at end 1. Similarly, n''_x, n''_y, and n''_z be mutually perpendicular unit vectors along the convenient oblique directions at end 2. Then define the elements of transformation matrices $[S']$ and $[S'']$ as follows:

$$S'_{k\ell} = n'_k \cdot n_\ell \,, \quad S''_{k\ell} = n''_k \cdot n_\ell \quad (k, \ell = x, y, z)$$

Then let transformation matrices T', T'', and T be defined as

$$[T'] = \left[\begin{array}{c|c} S' & 0 \\ \hline 0 & S' \end{array}\right], \quad [T''] = \left[\begin{array}{c|c} S'' & 0 \\ \hline 0 & S'' \end{array}\right], \quad [T] = \left[\begin{array}{c|c} T' & 0 \\ \hline 0 & T'' \end{array}\right]$$

(Show that $[T]$ is a 12×12 matrix.) Let $\{\hat{u}\} = [T]\{u\}$, $\{\hat{f}\} = [T]\{f\}$, and finally let $\{\hat{f}\} = [k]\{u\}$. Then show that $[\hat{k}] = [T]\,[k]\,[T]^T$.

4

Boundary Value Problems

4.1 INTRODUCTION

Chapters 2 and 3 serve as an introduction to the basic procedures of the finite element method. The emphasis has been on the element concept, the element stiffness matrix, local and global coordinates, and the assembly procedure. In this and the next chapter we will consider briefly more general and abstract approaches and applications of the finite element method. Specifically, in this chapter we will look at the application of the foregoing procedure in solving ordinary differential equations and boundary value problems. Our approach will be essentially the same as in the earlier chapters. That is, we will develop the ideas through example problems. Specifically, we will seek to use the finite element method to find the solution to the following boundary value problem: Find $y(x)$ such that

$$\frac{d^2y}{dx^2} - y = f(x) \qquad (4.1)$$

where

$$y(a) = c \quad \text{and} \quad y(b) = d \qquad (4.2)$$

In Chapter 1 we indicated that solving this problem is equivalent to finding a function $y(x)$ which minimizes the integral

$$I = \int_a^b [y^2 + (y')^2 + 2yf] \, dx \qquad (4.3)$$

subject to the boundary conditions of Equation (4.2).

In the following sections we will show that the minimizing function can be determined by dividing the interval $[a, b]$ into subintervals or elements and then by approximating $y(x)$ on these elements. We will then see that the resulting analysis is directly analogous to that of the truss and beam problems of Chapters 2 and 3. Hence, analogous solution procedures can be used.

4.2 ELEMENTS AND PYRAMID FUNCTIONS

To solve the boundary value problem of Equations (4.1) and (4.2) [or equivalently of Equation (4.2) and (4.3)] by the finite element method, it is convenient to introduce some new interpolation functions and some notation and conventions which will help us keep account of the details of the method. As in Chapters 2 and 3, none of these notations or conventions is particularly difficult or profound, but taken as a whole, they can be confusing to the casual reader. Therefore, we will attempt to be as careful as possible in our definitions and we will sometimes use several notations when it is felt they will help clarify or simplify the analysis.

To begin, consider the interval $[a, b]$ of the x-axis to be divided into R segments or elements as shown in Figure 4.1. For convenience, let the elements all have the same length.* Let the element

---

*This is only a convenience and is not necessary. Indeed, at times it may be more convenient to let the elements have different lengths.

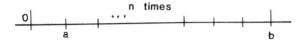

Figure 4.1 Elements of the Interval $[a, b]$

length be ℓ and let the coordinates of the ends of the elements be $X_0, X_1, \ldots, X_n$ as shown in Figure 4.2. Finally, let the elements themselves be numbered from 1 to n as also shown in Figure 4.2. Hence, we immediately have the following simple relations:

$$\ell = X_{e+1} - X_e = (b - a)/n, \quad e + 1 = 1, \ldots, n \qquad (4.4)$$

where X_e is the right-hand end or node of a typical element (e). That is, by following a notation similar to that of Chapters 2 and 3, we have

$$\overset{(e)}{x_1} = X_{e-1} \quad \text{and} \quad \overset{(e)}{x_2} = X_e \qquad (4.5)$$

where $\overset{(e)}{x_1}$ is the coordinate of the left-hand or "1" node of typical element (e) and $\overset{(e)}{x_2}$ is the right-hand or "2" node.

Next, let us introduce the so-called "pyramid" functions $\overset{(e)}{\phi}(x)$ as follows:

$$\overset{(e)}{\phi}(X) = \begin{cases} \dfrac{X - X_{e-1}}{X_e - X_{e-1}} = (X - X_{e-1})/\ell, & X_{e-1} \leqslant X \leqslant X_e \\[2ex] \dfrac{X_{e+1} - X}{X_{e+1} - X_e} = (X_{e+1} - X)/\ell, & X_e \leqslant X \leqslant X_{e+1} \\[2ex] 0, & X \leqslant X_{e-1} \quad \text{or} \quad X \geqslant X_{e+1} \end{cases} \qquad (4.6)$$

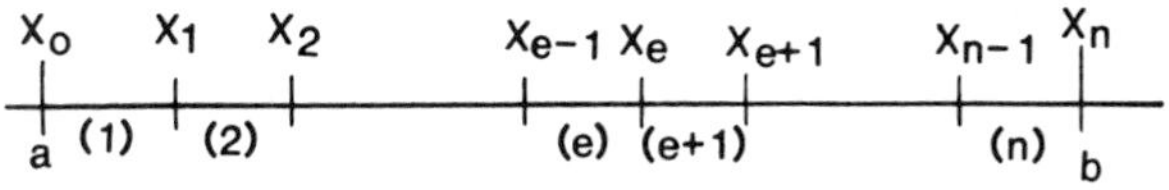

Figure 4.2 Numbering and Labeling of the Elements

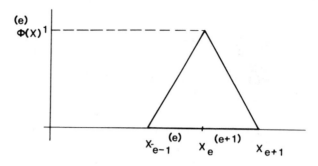

Figure 4.3 The Pyramid Function $\overset{(e)}{\phi}(X)$

A graph of $\overset{(e)}{\phi}(X)$ is shown in Figure 4.3. Note that $\overset{(e)}{\phi}(X)$ is nonzero over two elements, and that it has the value 1 at X_e; that is,

$$\overset{(e)}{\phi}(X_e) = 1 \qquad\qquad\qquad (4.7)$$

For convenience, we also define $\overset{(e)}{\phi}(X)$ to be zero outside the interval $[a, b]$. Hence, the graphs of $\overset{(0)}{\phi}(X)$ and $\overset{(n)}{\phi}(X)$ are as shown in Figure 4.4.

The pyramid functions are very useful in obtaining piecewise linear approximations to functions. To see this, consider, for example, dividing an interval into 4 elements. The the pyramid functions $\overset{(0)}{\phi}, \overset{(1)}{\phi}, \ldots, \overset{(4)}{\phi}$ would appear as shown in Figure 4.5. Now, suppose we want to obtain a piecewise linear approximation to some

Figure 4.4 Graphs of the Pyramid Functions $\overset{(0)}{\phi}(X)$ and $\overset{(n)}{\phi}(X)$

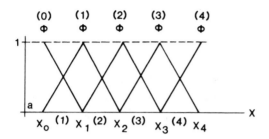

Figure 4.5 The Pyramid Functions $\overset{(0)}{\phi}, \overset{(1)}{\phi}, \dots, \overset{(4)}{\phi}$ for a 4-Element Interval Division

function $Y(X)$ over the interval $[a, b]$ as shown in Figure 4.6. Then, in terms of the pyramid functions this approximation takes the remarkably simple form

$$Y(X) = \sum_{e=0}^{4} Y_e \overset{(e)}{\phi}(X) \tag{4.8}$$

where Y_e ($e = 0, \dots, 4$) are the ordinates of $Y(X)$ at X_e ($e = 0, \dots, 4$) as shown in Figure 4.6. That is,

$$Y_e = Y(X_e) \tag{4.9}$$

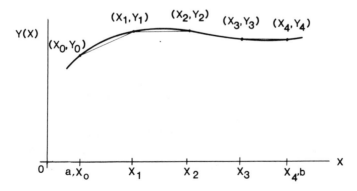

Figure 4.6 Piecewise Linear Approximation to a Function $Y(X)$

4.3   RELATION TO THE LINEAR
LAGRANGE FUNCTIONS

The pyramid functions may be related to the linear Lagrange interpolation functions described in Chapter 1. To see this, recall that if
(x_1, y_1) and (x_2, y_2) are the coordinates of two points in the X-Y
coordinate plane, the equation of the line passing through these
points is

$$y = N_1(x)y_1 + N_2(x)y_2 \qquad (4.10)$$

where $N_1(x)$ and $N_2(x)$ are the linear Lagrange functions given by
the expressions

$$N_1(x) = \frac{x - x_2}{x_1 - x_2} \quad \text{and} \quad N_2(x) = \frac{x - x_1}{x_2 - x_1} \qquad (4.11)$$

(see Chapter 1, Section 1.3).

To relate those functions to the pyramid functions and for our
later use, it is convenient to define the following *element* interpolation functions:

$$\overset{(e)}{N_1}(X) = \frac{X - \overset{(e)}{x_2}}{\overset{(e)}{x_1} - \overset{(e)}{x_2}} \quad \text{and} \quad \overset{(e)}{N_2}(X) = \frac{X - \overset{(e)}{x_1}}{\overset{(e)}{x_2} - \overset{(e)}{x_1}}$$

$$\text{for } \overset{(e)}{x_1} \leqslant X \leqslant \overset{(e)}{x_2} \qquad (4.12)$$

and

$$\overset{(e)}{N_1}(X) = \overset{(e)}{N_2}(X) = 0 \quad \text{for} \quad X < \overset{(e)}{x_1} \text{ or } X > \overset{(e)}{x_2} \qquad (4.13)$$

That is, the element interpolation functions are nonzero only on the
element itself. Figure 4.7 shows a graph of these functions. Hence,
by comparing Equations (4.6) and (4.12) and the graphs of Figures
4.3 and 4.7, it is seen that the pyramid functions $\overset{(e)}{\phi}(X)$ may be
written as

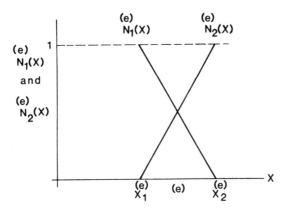

Figure 4.7   Element Interpolation Functions

$$\overset{(e)}{\phi}(X) = \begin{cases} \overset{(e)}{N_2}(X) & [\text{on element } (e)] \\ \overset{(e+1)}{N_1}(X) & [\text{on element } (e+1) \end{cases} \qquad (4.14)$$

Now, since from Equation (4.13) $\overset{(e)}{N_1}(X)$ and $\overset{(e)}{N_2}(X)$ are zero when x is not in element (e), Equation (4.14) may be written simply

$$\overset{(e)}{\phi}(X) = \overset{(e)}{N_2}(X) + \overset{(e+1)}{N_1}(X) \qquad (4.15)$$

To develop Equation (4.15) more formally, and again, for our later use, it is helpful at this point to introduce incidence matrices, similar to those of Chapters 2 and 3, as follows: Let the incidence matrix $[\overset{(e)}{\Lambda}]$ have $n + 1$ rows and 2 columns where n is the number of elements in the interval. Let the entries $\overset{(e)}{\lambda}_{ij}$ of $[\overset{(e)}{\Lambda}]$ be defined as follows:

$$\left[\overset{(1)}{\lambda}_{ij}\right] = \begin{bmatrix} 1 & 0 \\ 0 & 1 \\ 0 & 0 \\ 0 & 0 \\ \cdot & \cdot \\ \cdot & \cdot \\ \cdot & \cdot \\ 0 & 0 \\ 0 & 0 \end{bmatrix}, \qquad \left[\overset{(2)}{\lambda}_{ij}\right] = \begin{bmatrix} 0 & 0 \\ 1 & 0 \\ 0 & 1 \\ 0 & 0 \\ 0 & 0 \\ \cdot & \cdot \\ \cdot & \cdot \\ \cdot & \cdot \\ 0 & 0 \end{bmatrix}, \dots,$$

$$
\begin{bmatrix} (e) \\ \lambda_{ij} \end{bmatrix} = \begin{bmatrix} 0 & 0 \\ 0 & 0 \\ \cdot & \cdot \\ \cdot & \cdot \\ \cdot & \cdot \\ 1 & 0 \\ 0 & 1 \\ 0 & 0 \\ \cdot & \cdot \\ \cdot & \cdot \\ \cdot & \cdot \\ 0 & 0 \end{bmatrix} \leftarrow \text{row } e, \dots, \begin{bmatrix} (n) \\ \lambda_{ij} \end{bmatrix} = \begin{bmatrix} 0 & 0 \\ 0 & 0 \\ 0 & 0 \\ \cdot & \cdot \\ \cdot & \cdot \\ \cdot & \cdot \\ 0 & 0 \\ 1 & 0 \\ 0 & 1 \end{bmatrix} \begin{array}{l} \\ \\ \\ \\ \\ \\ \\ \leftarrow \text{row } n \\ \leftarrow \text{row } n+1 \end{array}
$$

$$(4.16)$$

Now, in terms of the $\overset{(e)}{\lambda}_{ij}$ it is easily seen that the pyramid functions are

$$
\overset{(k-1)}{\phi}(X) = \sum_{e=1}^{n} \sum_{j=1}^{2} \overset{(e)}{\lambda}_{kj} \overset{(e)}{N}_{j}(X)
\tag{4.17}
$$

or in matrix notation,

$$
\{\phi(X)\} = \sum_{e=1}^{n} \overset{(e)}{[\Lambda]} \{\overset{(e)}{N}(X)\}
\tag{4.18}
$$

where $\overset{(e)}{N}(X)$ is a column matrix with two entries $\overset{(e)}{N}_{1}(X)$ and $\overset{(e)}{N}_{2}(X)$, and $\{\phi(X)\}$ is a column matrix with $n+1$ entries $\overset{(k-1)}{\phi}(X)$ $(k = 1, \dots, n+1)$; that is,

$$
\{\overset{(e)}{N}(X)\} = \left\{ \begin{array}{c} \overset{(e)}{N}_{1}(X) \\ \overset{(e)}{N}_{2}(X) \end{array} \right\}
\tag{4.19}
$$

$$
\{\phi(X)\} = \left\{ \begin{array}{c} \overset{(0)}{\phi}(X) \\ \overset{(1)}{\phi}(X) \\ \cdot \\ \cdot \\ \overset{(n)}{\phi}(X) \end{array} \right\}
\tag{4.20}
$$

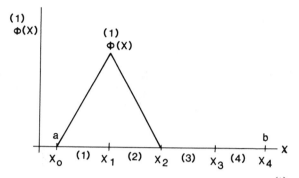

Figure 4.8 A Four Element Division of $[a, b]$ and $\overset{(1)}{\phi}(X)$

To illustrate this more explicitly, consider again, for example, a 4 element division of the interval $[a, b]$ as shown in Figures 4.5 and 4.6 and in Figure 4.8, where $\overset{(1)}{\phi}(X)$ is also shown. Then for example, from Equation (4.15) (or alternatively, by comparing Figures 4.7 and 4.8), we see that $\overset{(1)}{\phi}(X)$ is

$$\overset{(1)}{\phi}(X) = \overset{(1)}{N}_2(X) + \overset{(2)}{N}_1(X) \tag{4.21}$$

However, by expanding Equation (4.18) for $n = 4$, we find that $\Phi(X)$ is

$$\{\Phi(X)\} = \underbrace{\begin{bmatrix} 1 & 0 \\ 0 & 1 \\ 0 & 0 \\ 0 & 0 \\ 0 & 0 \end{bmatrix}}_{\overset{(1)}{\Lambda}} \begin{bmatrix} \overset{(1)}{N}_1 \\ \overset{(1)}{N}_2 \end{bmatrix} + \underbrace{\begin{bmatrix} 0 & 0 \\ 1 & 0 \\ 0 & 1 \\ 0 & 0 \\ 0 & 0 \end{bmatrix}}_{\overset{(2)}{\Lambda}} \begin{bmatrix} \overset{(2)}{N}_1 \\ \overset{(2)}{N}_2 \end{bmatrix}$$

$$+ \underbrace{\begin{bmatrix} 0 & 0 \\ 0 & 0 \\ 1 & 0 \\ 0 & 1 \\ 0 & 0 \end{bmatrix}}_{\overset{(3)}{\Lambda}} \begin{bmatrix} \overset{(3)}{N}_1 \\ \overset{(3)}{N}_2 \end{bmatrix} + \underbrace{\begin{bmatrix} 0 & 0 \\ 0 & 0 \\ 0 & 0 \\ 1 & 0 \\ 0 & 1 \end{bmatrix}}_{\overset{(4)}{\Lambda}} \begin{bmatrix} \overset{(4)}{N}_1 \\ \overset{(4)}{N}_2 \end{bmatrix} \tag{4.22}$$

or

$$\{\Phi(X)\} = \begin{Bmatrix} \overset{(1)}{N_1} \\ \overset{(1)}{N_2} \\ 0 \\ 0 \\ 0 \end{Bmatrix} + \begin{Bmatrix} 0 \\ \overset{(2)}{N_1} \\ \overset{(2)}{N_2} \\ 0 \\ 0 \end{Bmatrix} + \begin{Bmatrix} 0 \\ 0 \\ \overset{(3)}{N_1} \\ \overset{(3)}{N_2} \\ 0 \end{Bmatrix} + \begin{Bmatrix} 0 \\ 0 \\ 0 \\ \overset{(4)}{N_1} \\ \overset{(4)}{N_2} \end{Bmatrix} \qquad (4.23)$$

or finally,

$$\{\Phi(X)\} = \begin{Bmatrix} \overset{(1)}{N_1} \\ \overset{(1)}{N_2} + \overset{(2)}{N_1} \\ \overset{(2)}{N_2} + \overset{(3)}{N_1} \\ \overset{(3)}{N_2} + \overset{(4)}{N_1} \\ \overset{(4)}{N_2} \end{Bmatrix} \qquad (4.24)$$

Hence, as indicated in Equation (4.20), the pyramid functions $\overset{(k-1)}{\phi}(X)$ may be identified with a column matrix where the index k refers to the rows of the matrix. We then have the following relations [see also Equation (4.21) and Figure 4.4]:

$$\overset{(0)}{\phi}(X) = \overset{(1)}{N_1}(X)$$

$$\overset{(1)}{\phi}(X) = \overset{(1)}{N_2}(X) + \overset{(2)}{N_1}(X)$$

$$\overset{(2)}{\phi}(X) = \overset{(2)}{N_2}(X) + \overset{(3)}{N_1}(X) \qquad (4.25)$$

$$\overset{(3)}{\phi}(X) = \overset{(3)}{N_3}(X) + \overset{(4)}{N_1}(X)$$

$$\overset{(4)}{\phi}(X) = \overset{(4)}{N_2}$$

Finally, to verify Equation (4.15) analytically, note that in Equation (4.16), $\overset{(e)}{\lambda_{ij}}$ has a zero row unless i is either equal to e or $e + 1$ [see Equation (4.16)]. Moreover, in row i, $\overset{(e)}{\lambda_{ij}}$ has a nonzero 1st column entry only when i equals e and a nonzero 2nd column entry only when i equals $e + 1$. That is,

$$\overset{(e)}{\lambda_{ij}} = \delta_{pj} \quad \text{with} \quad p = i - e + 1 \tag{4.26}$$

where δ_{pj} is Kronecker's delta function with value one when $p = j$ and zero when $p \neq j$. By substituting from Equation (4.26) into Equation (4.17) we have

$$\overset{(k-1)}{\phi}(X) = \sum_{e=1}^{n} \sum_{j=1}^{2} \overset{(e)}{\lambda_{kj}} \overset{(e)}{N_j}(X) = \sum_{e=1}^{n} \sum_{j=1}^{2} \delta_{pj} \overset{(e)}{N_j}(X) \tag{4.27}$$

where p is $k - e + 1$. But since δ_{pj} is zero unless $p = j$, that is, unless $k - e + 1 = j$ or equivalently, unless $e = k - j + 1$, Equation (4.27) may be written

$$\overset{(k-1)}{\phi}(X) = \sum_{j=1}^{2} \overset{(k-j+1)}{N_j}(X) \tag{4.28}$$

By expanding the sum, we immediately obtain Equation (4.15). That is,

$$\overset{(k-1)}{\phi}(X) = \overset{(k)}{N_1}(X) + \overset{(k-1)}{N_2}(X)$$

or $\tag{4.29}$

$$\overset{(k)}{\phi}(X) = \overset{(k+1)}{N_1}(X) + \overset{(k)}{N_2}(X)$$

4.4 ELEMENT FUNCTION APPROXIMATION

In this section we will show how the element Lagrange functions can be conveniently used to obtain piecewise approximations—that is, "finite-element" approximations to functions.

Recall, in Section 4.2, we saw that in terms of the pyramid functions, a given function $Y(X)$ may be expressed as [see Equation (4.8)]

$$Y(X) = \sum_{k=0}^{n} Y_k \overset{(k)}{\phi}(X) \tag{4.30}$$

or equivalently, by adjusting the summation index, as

$$Y(X) = \sum_{k=1}^{n+1} Y_{k-1} \overset{(k-1)}{\phi}(X) \tag{4.31}$$

Hence, by substituting for $\overset{(k-1)}{\phi}(X)$ from Equation (4.17), we have

$$Y(X) = \sum_{k=1}^{n+1} Y_{k-1} \sum_{e=1}^{n} \sum_{j=1}^{2} \overset{(e)}{\lambda_{kj}} \overset{(e)}{N_j}(X) \tag{4.32}$$

or, by rearranging the summations,

$$Y(X) = \sum_{e=1}^{n} \sum_{j=1}^{2} \left(\sum_{k=1}^{n+1} Y_{k-1} \overset{(e)}{\lambda_{kj}} \right) \overset{(e)}{N_j}(X) \tag{4.33}$$

Consider the terms within the parentheses of Equation (4.33): Let these be labeled $\overset{(e)}{y_j}$. That is, let

$$\overset{(e)}{y_j} \overset{D}{=} \sum_{k=1}^{n+1} Y_{k-1} \overset{(e)}{\lambda_{kj}} \tag{4.34}$$

Then, it is immediately seen by Equation (4.16) that the $\overset{(e)}{y_j}$ $(j = 1, 2)$ are simply the ordinates at the "1" and "2" ends or nodes of element (e). That is,

$$\overset{(e)}{y_j} = \overset{(e)}{Y(x_j)} \tag{4.35}$$

Hence, Equation (4.33) may be written

$$Y(X) = \sum_{e=1}^{n} \sum_{j=1}^{2} \overset{(e)}{y_j} \overset{(e)}{N_j}(X) \tag{4.36}$$

In matrix form, this may be written

$$Y(X) = \sum_{e=1}^{n} \{ \overset{(e)}{y} \}^T \{ \overset{(e)}{N(X)} \} \tag{4.37}$$

Recall again that our objective is to obtain a solution to the boundary value problem of Equations (4.1) and (4.2) which is equivalent to minimizing the integral of Equation (4.3). Therefore, we need to obtain expressions not only for $Y(X)$, but also for $Y'(X)$, $Y^2(X)$, $[Y'(X)]^2$, and $Y(X)f(X)$. Let us consider first $Y'(X)$: From Equations (4.36) and (4.37), this is simply

$$Y'(X) = \sum_{e=1}^{n} \sum_{j=1}^{2} \overset{(e)}{y_j} \overset{(e)}{N_j'}(X) \qquad (4.38)$$

or in matrix form

$$Y'(X) = \sum_{e=1}^{n} \{\overset{(e)}{y}\}^T \{\overset{(e)}{N'}(X)\} \qquad (4.39)$$

From Equation (4.19) $\overset{(e)}{N'}(X)$ is seen to be

$$\{\overset{(e)}{N'}(X)\} = \left\{ \begin{array}{c} \overset{(e)}{N_1'}(X) \\ \overset{(e)}{N_2'}(X) \end{array} \right\} \qquad (4.40)$$

where $\overset{(e)}{N_1'}(X)$ and $\overset{(e)}{N_2'}(X)$ are found from Equations (4.12) and (4.4) to be

$$\overset{(e)}{N_1'}(X) = 1/(\overset{(e)}{x_1} - \overset{(e)}{x_2}) = -1/\ell \qquad (4.41)$$

and

$$\overset{(e)}{N_2'}(X) = 1/(\overset{(e)}{x_2} - \overset{(e)}{x_1}) = 1/\ell \qquad (4.42)$$

Consider next $Y^2(X)$. From Equation (4.36) we might at first suppose that $Y^2(X)$ is

$$\left[\sum_{e=1}^{n} \sum_{j=1}^{2} \overset{(e)}{y_j} \overset{(e)}{N_j}(X) \right]^2$$

But clearly, this is not the meaning of $Y^2(X)$. That is, $Y^2(X)$ refers to the square of $Y(X)$ for any value on the abscissa. Since Equation (4.36) represents a piecewise or element approximation of $Y(X)$, $Y^2(X)$ must be obtained in the same manner. Specifically, $Y^2(X)$ must also be piecewise represented over the respective elements.

Hence, the correct expression for $Y^2(X)$ is

$$Y^2(X) = \sum_{e=1}^{n} \left[\sum_{j=1}^{2} \overset{(e)}{y_j} \overset{(e)}{N_j(X)} \right]^2 \tag{4.43}$$

Or, alternatively, from Equation (4.37), $Y^2(X)$ may be written

$$Y^2(X) = \sum_{e=1}^{n} (\{\overset{(e)}{y}\}^T \{\overset{(e)}{N}\}\{\overset{(e)}{N}\}^T \{\overset{(e)}{y}\}) \tag{4.44}$$

Similarly, from Equation (4.38) it is seen that $[Y'(X)]^2$ is

$$[Y'(X)]^2 = \sum_{e=1}^{n} \left[\sum_{j=1}^{2} \overset{(e)}{y_j} \overset{(e)}{N'_j(X)} \right]^2 \tag{4.45}$$

Or, alternatively, from Equation (4.39), $Y'(X)^2$ is

$$[Y'(X)]^2 = \sum_{e=1}^{n} (\{\overset{(e)}{y}\}^T \{\overset{(e)}{N'}\}\{\overset{(e)}{N'}\}^T \{\overset{(e)}{y}\}) \tag{4.46}$$

Finally, let us consider $Y(X)f(X)$. In direct analogy to Equation (4.36), let $f(X)$ be expressed as

$$f(X) = \sum_{e=1}^{n} \sum_{j=1}^{2} \overset{(e)}{f_j} \overset{(e)}{N_j(X)} \tag{4.47}$$

In matrix form, this is

$$f(X) = \sum_{e=1}^{n} \{\overset{(e)}{f}\}^T \{\overset{(e)}{N(X)}\} \tag{4.48}$$

Then, $Y(X)f(X)$ becomes

$$Y(X)f(X) = \sum_{e=1}^{n} \left[\sum_{j=1}^{2} \overset{(e)}{y_j} \overset{(e)}{N_j(X)} \right] \left[\sum_{k=1}^{2} \overset{(e)}{f_k} \overset{(e)}{N_k(X)} \right] \tag{4.49}$$

Or, finally,

$$Y(X)f(X) = \sum_{e=1}^{n} (\{\overset{(e)}{y}\}^{T} \{\overset{(e)}{N}\} \{\overset{(e)}{N}\}^{T} \{\overset{(e)}{f}\})$$ (4.50)

4.5 GENERALIZATION

Equations (4.36) and (4.37) represent the principal results of the foregoing section. They show, for example, that once the element interpolation functions $\overset{(e)}{N_j}(x)$ are known, the function approximation follows immediately by specifying the values of the function at the ends or nodes of the respective elements. Moreover, the form of Equation (4.37) is general in that it holds for higher order, as well as linear, interpolation functions.

To illustrate this, consider the quadratic interpolation function: Recall that if (x_1, y_1), (x_2, y_2), and (x_3, y_3) are the coordinates of three points in the X-Y coordinate plane, then the equation of the parabola passing through these points is

$$y = N_1(x)y_1 + N_2(x)y_2 + N_3(x)y_3$$ (4.51)

where $N_1(x)$, $N_2(x)$, and $N_3(x)$ are now the quadratic Lagrange interpolation functions given by the expressions

$$N_1(x) = \frac{(x - x_2)(x - x_3)}{(x_1 - x_2)(x_1 - x_3)}$$

$$N_2(x) = \frac{(x - x_1)(x - x_3)}{(x_2 - x_1)(x_2 - x_3)}$$ (4.52)

$$N_3(x) = \frac{(x - x_1)(x - x_2)}{(x_3 - x_1)(x_3 - x_2)}$$

Consider again the interval $[a, b]$ of the X-axis and let the interval be divided into three elements as shown in Figure 4.9. Observe that in this partitioning of the interval each element now

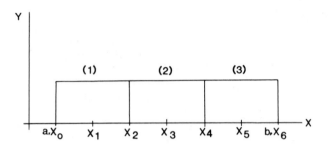

Figure 4.9 A Three-Element Division
of the X-Axis for Quadratic Interpolation

has three nodes and that there are a total of seven nodes. Hence, in general, for an n-element division there are a total of $2n + 1$ nodes.

Consider a typical element (e) of the interval as shown in Figure 4.10. By following the notation pattern of Section 4.3, the element Lagrange functions are:

$$\overset{(e)}{N_1}(X) = \frac{(X - \overset{(e)}{x_2})(X - \overset{(e)}{x_3})}{(\overset{(e)}{x_1} - \overset{(e)}{x_2})(\overset{(e)}{x_1} - \overset{(e)}{x_3})}$$

$$\overset{(e)}{N_2}(X) = \frac{(X - \overset{(e)}{x_1})(X - \overset{(e)}{x_3})}{(\overset{(e)}{x_2} - \overset{(e)}{x_1})(\overset{(e)}{x_2} - \overset{(e)}{x_3})} \tag{4.53}$$

$$\overset{(e)}{N_3}(X) = \frac{(X - \overset{(e)}{x_1})(X - \overset{(e)}{x_2})}{(\overset{(e)}{x_3} - \overset{(e)}{x_1})(\overset{(e)}{x_3} - \overset{(e)}{x_2})}$$

where $\overset{(e)}{x_1}$, $\overset{(e)}{x_2}$, and $\overset{(e)}{x_3}$ are the local, or element, node coordinates. Also, as in Section 4.3, let these element interpolation functions be zero outside the element. That is, as before, the element interpolation functions are nonzero only on the element itself. The graphs of $\overset{(e)}{N_1}(X)$, $\overset{(e)}{N_2}(X)$, and $\overset{(e)}{N_3}(X)$ are shown in Figure 4.10. It is easily shown that these functions have the property

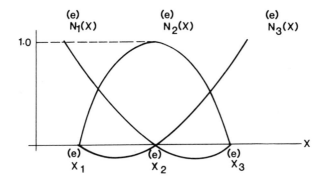

Figure 4.10 Graphical Representation of $\overset{(e)}{N_1}(X), \overset{(e)}{N_2}(X)$, and $\overset{(e)}{N_3}(X)$

$$\overset{(e)}{N_1}(X) + \overset{(e)}{N_2}(X) + \overset{(e)}{N_3}(X) = 1 \tag{4.54}$$

and that at the nodes $\overset{(e)}{x_1}, \overset{(e)}{x_2}$, and $\overset{(e)}{x_3}$ they have the values as shown in Table 4.1.

Consider now a "finite element" representation of a function $Y(X)$ over the interval $[a, b]$ as shown in Figure 4.11. Using Equation (4.51) such a representation is:

$$Y(X) = \underbrace{Y_0 \overset{(1)}{N_1} + Y_1 \overset{(1)}{N_2} + Y_2 \overset{(1)}{N_3}}_{\text{element (1)}} + \underbrace{Y_2 \overset{(2)}{N_1} + Y_3 \overset{(2)}{N_2} + Y_4 \overset{(2)}{N_3}}_{\text{element (2)}}$$

$$+ \underbrace{Y_4 \overset{(3)}{N_1} + Y_5 \overset{(3)}{N_2} + Y_6 \overset{(3)}{N_3}}_{\text{element (3)}} \tag{4.55}$$

Table 4.1 Nodes Values of $\overset{(e)}{N_i}(X)$
$(i = 1, 2, 3)$

	$\overset{(e)}{x_1}$	$\overset{(e)}{x_2}$	$\overset{(e)}{x_3}$
$\overset{(e)}{N_1}(X)$	1	0	0
$\overset{(e)}{N_2}(X)$	0	1	0
$\overset{(e)}{N_3}(X)$	0	0	1

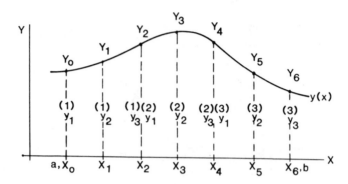

Figure 4.11 Quadratic Finite Element Representation of a Function

or

$$Y(X) = \sum_{e=1}^{3} \sum_{j=1}^{3} \overset{(e)}{y_j} \overset{(e)}{N_j}(X) \tag{4.56}$$

or

$$Y(X) = \sum_{e=1}^{3} \{\overset{(e)}{y}\}^T \{\overset{(e)}{N}(X)\} \tag{4.57}$$

where $\overset{(e)}{y_j}$ $(e, j = 1, 2, 3)$ are the element ordinates of $Y(X)$ as shown in Figure 4.11.

Equations (4.56) and (4.57) are therefore seen to have exactly the same form as Equations (4.36) and (4.37).

4.6 GLOBAL COORDINATES

Another expression developed in Section 4.4 which merits additional comment is Equation (4.34), that is,

$$\overset{(e)}{y_j} = \sum_{k=1}^{n+1} Y_{k-1} \overset{(e)}{\lambda_{kj}} \tag{4.58}$$

This expression is significant in that it provides a relationship between the local coordinates $\overset{(e)}{y_j}$ and the global coordinates y_j.

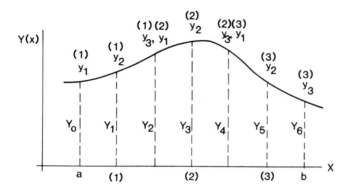

Figure 4.12 Local and Global Coordinates for a Quadratic
Interpolation of a 3-Element Interval Partitioning

Recall that in Chapters 2 and 3 the relationships between the local
and global coordinates were via incidence matrices. The same is the
case here.

To explore and illustrate the application of Equation (4.58),
consider again the quadratic interpolation over the elements of a
3-element partitioning of an interval as in Figures 4.9, 4.10, and
4.11, and as shown in Figure 4.12. By inspection of Figure 4.12,
it is seen that

$$Y_0 = \overset{(1)}{y}_1$$

$$Y_1 = \overset{(1)}{y}_2$$

$$Y_2 = \overset{(1)}{y}_3 = \overset{(2)}{y}_1$$

$$Y_3 = \overset{(2)}{y}_2 \qquad\qquad (4.59)$$

$$Y_4 = \overset{(2)}{y}_3 = \overset{(3)}{y}_1$$

$$Y_5 = \overset{(3)}{y}_2$$

$$Y_6 = \overset{(3)}{y}_3$$

To check that these relations are consistent with Equation (4.58), let us introduce incidence matrices in the same manner as in Section 4.3 in Equation (4.16). That is, let the matrices $[\overset{(1)}{\Lambda}]$, $[\overset{(2)}{\Lambda}]$, and $[\overset{(3)}{\Lambda}]$ be

$$
[\overset{(1)}{\Lambda}] = \begin{bmatrix} 1 & 0 & 0 \\ 0 & 1 & 0 \\ 0 & 0 & 1 \\ 0 & 0 & 0 \\ 0 & 0 & 0 \\ 0 & 0 & 0 \\ 0 & 0 & 0 \end{bmatrix}, \quad
[\overset{(2)}{\Lambda}] = \begin{bmatrix} 0 & 0 & 0 \\ 0 & 0 & 0 \\ 1 & 0 & 0 \\ 0 & 1 & 0 \\ 0 & 0 & 1 \\ 0 & 0 & 0 \\ 0 & 0 & 0 \end{bmatrix}, \quad
[\overset{(3)}{\Lambda}] = \begin{bmatrix} 0 & 0 & 0 \\ 0 & 0 & 0 \\ 0 & 0 & 0 \\ 0 & 0 & 0 \\ 1 & 0 & 0 \\ 0 & 1 & 0 \\ 0 & 0 & 1 \end{bmatrix}
$$

$$(4.60)$$

Next, as in Chapters 2 and 3, let the global coordinate array $\{Y\}$ be defined as

$$
\{Y\} = \begin{bmatrix} Y_0 \\ Y_1 \\ Y_2 \\ Y_3 \\ Y_4 \\ Y_5 \\ Y_6 \end{bmatrix}
$$

$$(4.61)$$

Then, in terms of $\{Y\}$, Equation (4.58) may be written in the form

$$
\{\overset{(e)}{y}\}^T = \{Y\}^T [\overset{(e)}{\Lambda}]
$$

$$(4.62)$$

or, equivalently as

$$
\{\overset{(e)}{y}\} = [\overset{(e)}{\Lambda}]^T \{Y\}
$$

$$(4.63)$$

where the $\overset{(e)}{\lambda}_{ij}$ of Equation (4.58) are the entries of the $[\overset{(e)}{\Lambda}]$ incidence matrices. An inspection of Equations (4.58) through (4.63) shows that Equations (4.58) and (4.59) are indeed consistent.

As in Chapters 2 and 3, Equations (4.62) and (4.63) will prove to be useful in the sequel.

4.7   STIFFNESS MATRICES

Let us now return again to the boundary value problem which we used to introduce the ideas of this chapter. Recall that [see Section 4.1, Equation (4.3)] we were seeking a function $y(x)$ which would minimize the integral

$$I = \int_a^b [y^2 + (y')^2 + 2yf]\, dx \tag{4.66}$$

We can now make a "finite element" approximation to $y(x)$ by using the piecewise approximations developed in Section 4.4. Specifically, by using Equations (4.44), (4.46), and (4.50), we can write I in the form

$$I = \int_a^b \left[\sum_{e=1}^n \left\langle \{\overset{(e)}{y}\}^T\{\overset{(e)}{N}\} \, \{\overset{(e)}{N}\}^T\{\overset{(e)}{y}\} + \{\overset{(e)}{y}\}^T\{\overset{(e)}{N'}\} \, \{\overset{(e)}{N'}\}^T\{\overset{(e)}{y}\} \right. \right.$$
$$\left. \left. + 2\{\overset{(e)}{y}\}^T\{\overset{(e)}{N}\}\{\overset{(e)}{N}\}^T\{\overset{(e)}{f}\} \right\rangle \right] dx \tag{4.67}$$

By noting that the summation and integration can be interchanged, and by remembering that y and f are independent of x, I may be written in the form

$$I = \sum_{e=1}^n \overset{(e)}{I} \tag{4.68}$$

where $\overset{(e)}{I}$, the "element integral" is

$$\overset{(e)}{I} = \{\overset{(e)}{y}\}^T[\overset{(e)}{k}]\{\overset{(e)}{y}\} + 2\{\overset{(e)}{y}\}^T\{\overset{(e)}{g}\} \tag{4.69}$$

where $[\overset{(e)}{k}]$ and $\{\overset{(e)}{g}\}$ are defined as

$$[\overset{(e)}{k}] = \int_a^b [\{\overset{(e)}{N}\}\{\overset{(e)}{N}\}^T + \{\overset{(e)}{N}\}'\{\overset{(e)}{N}\}'^T]\, dx \tag{4.70}$$

and

$$\overset{(e)}{g} = \left(\int_a^b [\{\overset{(e)}{N}\}\{\overset{(e)}{N}\}^T]\, dx \right)\{\overset{(e)}{f}\} \tag{4.71}$$

As in Chapters 2 and 3, $[\overset{(e)}{k}]$ is called the "local" or "element stiffness matrix," and $\{\overset{(e)}{g}\}$ is called the "local" or "element forcing array." (Note that, as with the previous stiffness matrices, $[\overset{(e)}{k}]$ is symmetrical.)

The analogous "global" stiffness matrix can be obtained by using Equations (4.62) and (4.63) of the previous section, which expresses the local $\{\overset{(e)}{y}\}$ array in terms of the global $\{Y\}$ array. Specifically, by using these expressions, I may be written in the form

$$I = \{Y\}^T [K]\{Y\} + 2\{Y\}^T \{G\} \tag{4.72}$$

where $[K]$ and $\{G\}$ are

$$[K] = \sum_{e=1}^{n} [\overset{(e)}{\Lambda}][\overset{(e)}{k}][\overset{(e)}{\Lambda}]^T \tag{4.73}$$

and

$$\{G\} = \sum_{e=1}^{n} [\overset{(e)}{\Lambda}]\{\overset{(e)}{g}\} = \left\langle \sum_{e=1}^{n} [\overset{(e)}{\Lambda}] \left(\int_a^b \{\overset{(e)}{N}\} \{\overset{(e)}{N}\}^T dx \right) [\Lambda]^T \right\rangle \{F\} \tag{4.74}$$

where $\{F\}$ is the global representation of $f(x)$. That is, from Equation (4.63) $\{\overset{(e)}{f}\}$ and $\{F\}$ are related by the expression

$$\{\overset{(e)}{f}\} = [\overset{(e)}{\Lambda}]^T \{F\} \tag{4.75}$$

$[K]$ and $\{G\}$ are called the "global stiffness matrix" and "global forcing array," respectively.

In our example boundary value problem, let us divide the interval $[a, b]$ into four elements as in Figure 4.13, and let us use the linear interpolation functions of Equation (4.12), that is, let $\{\overset{(e)}{N}\}$ be

$$\{\overset{(e)}{N}\} = \left\{ \begin{array}{c} \overset{(e)}{N_1}(x) \\ \overset{(e)}{N_2}(x) \end{array} \right\} \tag{4.76}$$

where $\overset{(e)}{N_1}(x)$ and $\overset{(e)}{N_2}(x)$ are

$$\overset{(e)}{N_1}(x) = \frac{x - \overset{(e)}{x_2}}{\overset{(e)}{x_1} - \overset{(e)}{x_2}} \quad \text{and} \quad N_2(x) = \frac{x - \overset{(e)}{x_1}}{\overset{(e)}{x_2} - \overset{(e)}{x_1}} \quad \text{for } \overset{(e)}{x_1} \leqslant x \leqslant \overset{(e)}{x_2} \tag{4.77}$$

Then the element stiffness array becomes

$$\overset{(e)}{[k]} = \int_a^b [\{\overset{(e)}{N}\}\{\overset{(e)}{N}\}^T + \{\overset{(e)}{N'}\}\{\overset{(e)}{N'}\}^T] \, dx$$

$$= \int_{\overset{(e)}{x_1}}^{\overset{(e)}{x_2}} [\{\overset{(e)}{N}\}\{\overset{(e)}{N}\} + \{\overset{(e)}{N'}\}\{\overset{(e)}{N'}\}^T] \, dx$$

$$= \int_{\overset{(e)}{x_1}}^{\overset{(e)}{x_2}} \left[\left\{ \begin{array}{c} \overset{(e)}{N_1}(x) \\ \overset{(e)}{N_2}(x) \end{array} \right\} \{\overset{(e)}{N_1}(x)\ \overset{(e)}{N_2}(x)\} \right.$$

$$\left. + \left\{ \begin{array}{c} N_1'(x) \\ N_2'(x) \end{array} \right\} \{\overset{(e)}{N_1}(x)\ \overset{(e)}{N_2}(x)\} \right] dx$$

$$= \int_{\overset{(e)}{x_1}}^{\overset{(e)}{x_2}} \left[\begin{array}{cc} (\overset{(e)}{N_1}\overset{(e)}{N_1} + \overset{(e)}{N_1'}\overset{(e)}{N_1'}) & (\overset{(e)}{N_1}\overset{(e)}{N_2} + \overset{(e)}{N_1'}\overset{(e)}{N_2'}) \\ (\overset{(e)}{N_2}\overset{(e)}{N_1} + \overset{(e)}{N_2'}\overset{(e)}{N_1'}) & (\overset{(e)}{N_2}\overset{(e)}{N_2} + \overset{(e)}{N_2'}\overset{(e)}{N_2'}) \end{array} \right] dx \tag{4.78}$$

Using Equation (4.77) for $N_1(x)$ and $N_2(x)$, this becomes

$$\overset{(e)}{[k]} = \left[\begin{array}{cc} (\ell/3 + 1/\ell) & (\ell/6 - 1/\ell) \\ (\ell/6 - 1/\ell) & (\ell/3 + 1/\ell) \end{array} \right] \tag{4.79}$$

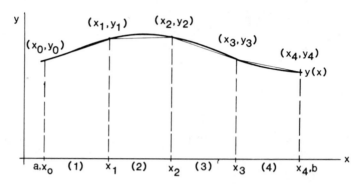

Figure 4.13 A 4-Element Partitioning of the Interval $[a, b]$
and A Piecewise Linear Approximation to $y(x)$

The global stiffness matrix is then [see Equations (4.16) and (4.73)]

$$
[K] = \begin{bmatrix} k_{11}^{(1)} & k_{12}^{(1)} & 0 & 0 & 0 \\ k_{21}^{(1)} & k_{22}^{(1)} & 0 & 0 & 0 \\ 0 & 0 & 0 & 0 & 0 \\ 0 & 0 & 0 & 0 & 0 \\ 0 & 0 & 0 & 0 & 0 \end{bmatrix} + \begin{bmatrix} 0 & 0 & 0 & 0 & 0 \\ 0 & k_{11}^{(2)} & k_{12}^{(2)} & 0 & 0 \\ 0 & k_{21}^{(2)} & k_{22}^{(2)} & 0 & 0 \\ 0 & 0 & 0 & 0 & 0 \\ 0 & 0 & 0 & 0 & 0 \end{bmatrix}
$$

$$
+ \begin{bmatrix} 0 & 0 & 0 & 0 & 0 \\ 0 & 0 & 0 & 0 & 0 \\ 0 & 0 & k_{11}^{(3)} & k_{12}^{(3)} & 0 \\ 0 & 0 & k_{21}^{(3)} & k_{22}^{(3)} & 0 \\ 0 & 0 & 0 & 0 & 0 \end{bmatrix} + \begin{bmatrix} 0 & 0 & 0 & 0 & 0 \\ 0 & 0 & 0 & 0 & 0 \\ 0 & 0 & 0 & 0 & 0 \\ 0 & 0 & 0 & k_{11}^{(4)} & k_{12}^{(4)} \\ 0 & 0 & 0 & k_{21}^{(4)} & k_{22}^{(4)} \end{bmatrix}
$$

$$
= \begin{bmatrix} k_{11}^{(1)} & k_{12}^{(1)} & 0 & 0 & 0 \\ k_{21}^{(1)} & (k_{22}^{(1)} + k_{11}^{(2)}) & k_{12}^{(2)} & 0 & 0 \\ 0 & k_{21}^{(2)} & (k_{22}^{(2)} + k_{11}^{(3)}) & k_{12}^{(3)} & 0 \\ 0 & 0 & k_{21}^{(3)} & (k_{22}^{(3)} + k_{11}^{(4)}) & k_{12}^{(4)} \\ 0 & 0 & 0 & k_{21}^{(4)} & k_{22}^{(4)} \end{bmatrix} \quad (4.80)
$$

Hence, from Equation (4.79), $[K]$ becomes

$$[K] = \begin{bmatrix} (\ell/3 + 1/\ell) & (\ell/6 - 1/\ell) & 0 & 0 & 0 \\ (\ell/6 - 1/\ell) & (2\ell/3 + 2/\ell) & (\ell/6 - 1/\ell) & 0 & 0 \\ 0 & (\ell/6 - 1/\ell) & (2\ell/3 + 2/\ell) & (\ell/6 - 1/\ell) & 0 \\ 0 & 0 & (\ell/6 - 1/\ell) & (2\ell/3 + 2/\ell) & (\ell/6 - 1/\ell) \\ 0 & 0 & 0 & (\ell/6 - 1/\ell) & (\ell/3 + 1/\ell) \end{bmatrix}$$

$$(4.81)$$

Similarly, using Equations (4.71) and (4.77), $\{g\}^{(e)}$ may be written (see Problem 4.12)

$$\{g\}^{(e)} = \begin{bmatrix} \ell/3 & \ell/6 \\ \ell/6 & \ell/3 \end{bmatrix} \begin{Bmatrix} f_1^{(e)} \\ f_2^{(e)} \end{Bmatrix}$$

$$(4.82)$$

Hence, by using Equation (4.74), the global force array becomes (see Problem 4.13)

$$\{G\} = \begin{bmatrix} \ell/3 & \ell/6 & 0 & 0 & 0 \\ \ell/6 & 2\ell/3 & \ell/6 & 0 & 0 \\ 0 & \ell/6 & 2\ell/3 & \ell/6 & 0 \\ 0 & 0 & \ell/6 & 2\ell/3 & \ell/6 \\ 0 & 0 & 0 & \ell/6 & \ell/3 \end{bmatrix} \begin{Bmatrix} F_1 \\ F_2 \\ F_3 \\ F_4 \\ F_5 \end{Bmatrix}$$

$$(4.83)$$

4.8 SOLUTION TO THE EXAMPLE PROBLEM

The expressions of Section 4.7 are directly applicable in the solution of our boundary value problem of Section 4.1. To see this, observe that in terms of the entries of the stiffness and force arrays, the integral I of Equations (4.3) and (4.66) becomes

$$I = \sum_{i=0}^{4} \sum_{j=0}^{4} K_{i+j,\,j+1} \; Y_i Y_j + 2 \sum_{i=0}^{4} Y_i G_i \tag{4.84}$$

where, as before, the interval is partitioned into 4 elements. (Regarding notation, the subscripts of K and Y differ by 1 since the first entry of Y_i is Y_0.) To minimize I with respect to the Y_i ($i = 0, \ldots, 4$), we simply set the partial derivatives of I with respect to each of the Y_i equal to zero. This immediately leads to the equations

$$\frac{\partial I}{\partial Y_i} = 0 = \sum_{j=0}^{4} K_{i+1,\,j+1} \; Y_j + G_i \quad (i = 0, 1, \ldots, 4) \tag{4.85}$$

In matrix form, these equations may be written

$$[K]\{Y\} = -\{G\} \tag{4.86}$$

Knowing $[K]$ and $\{G\}$, Equation (4.85) or (4.86) can be solved for $\{Y\}$ as an approximate, that is, "finite element" solution to the boundary value problem of Equations (4.1) and (4.2). To illustrate this more specifically, let us suppose, for simplicity, that the interval $[a, b]$ is $(0, 4)$, that $f(x)$ is a constant f_0, and that c and d are $1 - f_0$ and $e^4 - f_0$, respectively. That is, we will attempt to solve the problem

$$\frac{d^2 y}{dx^2} - y = f_0 \tag{4.87}$$

where

$$y(0) = 1 - f_0 \quad \text{and} \quad y(4) = e^4 - f_0 \tag{4.88}$$

With these conditions, it is easily shown (Problem 4.14) that the exact solution is

$$y(x) = e^x - f_0 \tag{4.89}$$

When the interval $[0, 4]$ is divided into 4 subintervals or elements, each with length 1, then in Equations (4.79) to (4.83), ℓ is 1 and the element or local stiffness matrix $\overset{(e)}{[k]}$ is [Equation (4.79)]

$$
\overset{(e)}{[k]} = \frac{1}{6}
\begin{bmatrix}
8 & -5 \\
-5 & 8
\end{bmatrix}
\tag{4.90}
$$

Then, from Equation (4.81), the global stiffness matrix $[K]$ is

$$
[K] = \frac{1}{6}
\begin{bmatrix}
8 & -5 & 0 & 0 & 0 \\
-5 & 16 & -5 & 0 & 0 \\
0 & -5 & 16 & -5 & 0 \\
0 & 0 & -5 & 16 & -5 \\
0 & 0 & 0 & -5 & 8
\end{bmatrix}
\tag{4.91}
$$

Similarly, using Equation (4.82), the element force array $\{g_e\}$ becomes

$$
\{g_e\} =
\begin{bmatrix}
1/3 & 1/6 \\
1/6 & 1/3
\end{bmatrix}
\begin{bmatrix}
f_0 \\
f_0
\end{bmatrix}
\tag{4.92}
$$

From Equation (4.83), the global force array $\{G\}$ becomes

$$
\{G\} = f_0
\begin{Bmatrix}
1/2 \\
1 \\
1 \\
1 \\
1/2
\end{Bmatrix}
\tag{4.93}
$$

Hence, the governing matrix equation [Equation (4.86)] becomes

$$\frac{1}{6}\begin{bmatrix} 8 & -5 & 0 & 0 & 0 \\ -5 & 16 & -5 & 0 & 0 \\ 0 & -5 & 16 & -5 & 0 \\ 0 & 0 & -5 & 16 & -5 \\ 0 & 0 & 0 & -5 & 8 \end{bmatrix} \begin{Bmatrix} Y_0 \\ Y_1 \\ Y_2 \\ Y_3 \\ Y_4 \end{Bmatrix} = -f_0 \begin{Bmatrix} 1/2 \\ 1 \\ 1 \\ 1 \\ 1/2 \end{Bmatrix} \qquad (4.94)$$

Equation (4.94) is the replacement or finite element representation of the differential equation of Equation (4.87). The boundary conditions of Equation (4.88) become

$$Y_0 = 1 - f_0 \quad \text{and} \quad Y_4 = e^4 - f_0 \qquad (4.95)$$

At this point, it should be noted that the system of Equation (4.94) subject to the boundary conditions of Equation (4.95) is overdetermined. That is, we have the equivalent of 7 equations for the 5 unknown Y_i ($i = 0, \ldots, 4$). To resolve this dilemma, recall that the procedure for solving a nonhomogeneous differential equation such as Equation (4.87) (see Problem 4.14) is to superpose the solution to the corresponding homogeneous equation (right-hand side zero) with a particular solution of the nonhomogeneous equation. The boundary conditions are then used to determine the integration constants occurring in the homogeneous equation solution. Hence, let the solution of our system of Equations (4.94) and (4.95) take the form

$$\{Y\} = \{Y_H\} + \{Y_p\} \qquad (4.96)$$

where $\{Y_H\}$ is a solution to Equation (4.94) with f_0 being zero, and $\{Y_p\}$ is a "particular" solution of Equation (4.94). That is, let $\{Y_H\}$ and $\{Y_p\}$ satisfy the systems

$$\frac{1}{6}\begin{bmatrix} 8 & -5 & 0 & 0 & 0 \\ -5 & 16 & -5 & 0 & 0 \\ 0 & -5 & 16 & -5 & 0 \\ 0 & 0 & -5 & 16 & -5 \\ 0 & 0 & 0 & -5 & 8 \end{bmatrix} \begin{Bmatrix} Y_{H0} \\ Y_{H1} \\ Y_{H2} \\ Y_{H3} \\ Y_{H4} \end{Bmatrix} = 0 \qquad (4.97)$$

and

$$\frac{1}{6} \begin{bmatrix} 8 & -5 & 0 & 0 & 0 \\ -5 & 16 & -5 & 0 & 0 \\ 0 & -5 & 16 & -5 & 0 \\ 0 & 0 & -5 & 16 & -5 \\ 0 & 0 & 0 & -5 & 8 \end{bmatrix} \begin{Bmatrix} Y_{P0} \\ Y_{P1} \\ Y_{P2} \\ Y_{P3} \\ Y_{P4} \end{Bmatrix} = -f_0 \begin{Bmatrix} 1/2 \\ 1 \\ 1 \\ 1 \\ 1/2 \end{Bmatrix} \tag{4.98}$$

Since the boundary conditions of a differential equation are satisfied by adjusting the values of constants arising in the homogeneous equation solution, let us seek to satisfy the boundary conditions of Equation (4.95) through the homogeneous equation solution $\{Y_H\}$. Now, since the boundary conditions of Equation (4.95) specify the values of Y_0 and Y_4, and since the governing system of Equation (4.86) is determined by varying each of the Y_i [Equation (4.85)], it is necessary to modify Equation (4.97) to the form

$$\frac{1}{6} \begin{bmatrix} -5 & 16 & -5 & 0 & 0 \\ 0 & -5 & 16 & -5 & 0 \\ 0 & 0 & -5 & 16 & -5 \end{bmatrix} \begin{Bmatrix} Y_{H0} \\ Y_{H1} \\ Y_{H2} \\ Y_{H3} \\ Y_{H4} \end{Bmatrix} = 0 \tag{4.99}$$

That is, the first and fifth rows, or equations, are deleted since Y_{H0} and Y_{H4} are to be determined from the boundary conditions.

Equation (4.99) is seen to be equivalent to the system

$$16Y_{H1} - 5Y_{H2} = 5Y_{H0}$$

$$-5Y_{H1} + 16Y_{H2} - 5Y_{H3} = 0 \tag{4.100}$$

$$-5Y_{H2} + 16Y_{H3} = 5Y_{H4}$$

Solving Equations (4.100) for Y_{H1}, Y_{H2}, and Y_{H3}, we obtain

$$Y_{H1} = (1155/3296)\, Y_{H0} + (125/3296)\, Y_{H4}$$

$$Y_{H2} = (25/206)\, Y_{H0} + (25/206)\, Y_{H4} \qquad (4.101)$$

$$Y_{H3} = (125/3296)\, Y_{H0} + (1155/3296)\, Y_{H4}$$

Similarly, Equation (4.98) is seen to be equivalent to the system

$$8Y_{P0} - 5Y_{P1} = -3f_0$$

$$-5Y_{P0} + 16Y_{P1} - 5Y_{P2} = -6f_0$$

$$-5Y_{P1} + 16Y_{P2} - 5Y_{P3} = -6f_0 \qquad (4.102)$$

$$-5Y_{P2} + 16Y_{P3} - 5Y_{P4} = -6f_0$$

$$-5Y_{P3} + 8Y_{P4} = -3f_0$$

Solving Equations (4.102) for Y_{Pi} ($i = 0, 1, \ldots, 4$) leads to the simple results

$$Y_{P0} = Y_{P1} = Y_{P2} = Y_{P3} = Y_{P4} = -f_0 \qquad (4.103)$$

Therefore, by combining the results of Equations (4.101) and (4.103) [see Equation (4.96)], the finite element solution is

$$\{Y\} = \{Y_H\} + \{Y_P\} = \begin{Bmatrix} Y_{H0} \\ (1155/3296)\, Y_{H0} + (125/3296)\, Y_{H4} \\ (25/206)\, Y_{H0} + (25/206)\, Y_{H4} \\ (125/3296)\, Y_{H0} + (1155/3296)\, Y_{H4} \\ Y_{H4} \end{Bmatrix} + \begin{Bmatrix} -f_0 \\ -f_0 \\ -f_0 \\ -f_0 \\ -f_0 \end{Bmatrix}$$

$$(4.104)$$

The boundary conditions of Equation (4.95) are then satisfied by taking Y_{H0} and Y_{H4} to be

$$Y_{H0} = 1 \tag{4.105}$$

and

$$Y_{H4} = e^4 \tag{4.106}$$

Hence, the solution becomes

$$\{Y\} = \begin{Bmatrix} -f_0 + 1.0 \\ -f_0 + 2.421 \\ -f_0 + 6.747 \\ -f_0 + 19.170 \\ -f_0 + 54.598 \end{Bmatrix} \tag{4.107}$$

The solution of Equation (4.87) with the boundary conditions of Equation (4.88) at the points $x = 0, 1, 2, 3, 4$ is [see Equation (4.89)]

$$Y = \begin{Bmatrix} -f_0 + 1.0 \\ -f_0 + 2.718 \\ -f_0 + 7.389 \\ -f_0 + 20.086 \\ -f_0 + 54.598 \end{Bmatrix} \tag{4.108}$$

A comparison of Equations (4.107) and (4.108) shows a very close agreement.

4.9   DISCUSSION

Equation (4.86) is of the same general form as the force-displacement equation we encountered in the finite element solution of the truss and beam problems. However, as with the truss and beam solutions, the actual computation involved in solving even a simple problem is excessive. Hence, the practical approach with these problems is to automate the procedure by developing algorithms for a computer program. As mentioned earlier, such algorithms have already been developed and an increasing number of computer programs to solve these problems are being written. It is beyond the scope of this text to discuss these algorithms or programs. However, the reader is encouraged to investigate and/or develop these algorithms according to his or her own interests.

4.10   FURTHER GENERALIZATION:
HERMITIAN INTERPOLATES

The foregoing procedures are an attempt to approximate a function by passing polynomials through selected points of the graph of the function. This leads to an approximation of the function which is continuous at all points, although the derivatives may be discontinuous at the nodes. In most cases, this is a reasonable and convenient

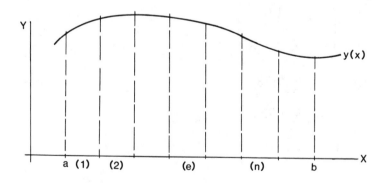

Figure 4.14   Element Division for a Hermitian
Approximation of a Function

approach. However, in some instances it may be desirable to have a smoother, or more conforming, approximation which is not only continuous, but which also has continuous derivatives at the nodes. Such an approximation can be obtained by using the Hermite polynomials introduced in Chapter 1.

To illustrate this, let us consider again the problem of approximating a function over an interval as shown in Figure 4.14. Consider a typical element (e) of the interval. Let us introduce on (e) the following element Hermitian interpolation functions (see Section 1.5)

$$\overset{(e)}{N_1}(x) = (x - \overset{(e)}{x_2})^2 \, [(\overset{(e)}{x_1} - \overset{(e)}{x_2}) + 2(\overset{(e)}{x_1} - x)] / (\overset{(e)}{x_1} - \overset{(e)}{x_2})^3$$

$$\overset{(e)}{N_2}(x) = (x - \overset{(e)}{x_1}) \, (x - \overset{(e)}{x_2})^2 / (\overset{(e)}{x_1} - \overset{(e)}{x_2})^2$$

$$\overset{(e)}{N_3}(x) = (x - \overset{(e)}{x_1})^2 \, [(\overset{(e)}{x_2} - \overset{(e)}{x_1}) + 2(\overset{(e)}{x_2} - x)] / (\overset{(e)}{x_2} - \overset{(e)}{x_1})^3 \quad (4.109)$$

$$\overset{(e)}{N_4}(x) = (x - \overset{(e)}{x_1})^2 \, (x - \overset{(e)}{x_2}) / (\overset{(e)}{x_1} - \overset{(e)}{x_2})^2$$

for $\overset{(e)}{x_1} \leqslant x \leqslant \overset{(e)}{x_2}$

Further, let these functions be zero outside the element (e); that is, let

$$\overset{(e)}{N_1}(x) = \overset{(e)}{N_2}(x) = \overset{(e)}{N_3}(x) = \overset{(e)}{N_4}(x) = 0 \quad (4.110)$$

for $x < \overset{(e)}{x_1}$ or $x > \overset{(e)}{x_2}$

Notice that these functions, as defined in Equation (4.109), also have the properties

$$\overset{(e)}{N_1}(\overset{(e)}{x_1}) = 1, \; \overset{(e)}{N_1}(\overset{(e)}{x_2}) = \overset{(e)}{N_1'}(\overset{(e)}{x_1}) = \overset{(e)}{N_1'}(\overset{(e)}{x_2}) = 0$$

$$\overset{(e)}{N_2'}(\overset{(e)}{x_1}) = 1, \; \overset{(e)}{N_2}(\overset{(e)}{x_1}) = \overset{(e)}{N_2}(\overset{(e)}{x_2}) = \overset{(e)}{N_2'}(\overset{(e)}{x_2}) = 0$$

$$\overset{(e)}{N_3}(\overset{(e)}{x_2}) = 1, \; \overset{(e)}{N_3}(\overset{(e)}{x_1}) = \overset{(e)}{N_3'}(\overset{(e)}{x_1}) = \overset{(e)}{N_3'}(\overset{(e)}{x_2}) = 0 \quad (4.111)$$

$$\overset{(e)}{N_4'}(\overset{(e)}{x_2}) = 1, \; \overset{(e)}{N_4}(\overset{(e)}{x_1}) = \overset{(e)}{N_4}(\overset{(e)}{x_2}) = \overset{(e)}{N_4'}(\overset{(e)}{x_1}) = 0$$

Suppose the values of the function $y(x)$ and its first derivatives are given at the nodes of the element. Then, in the interior of the element, $y(x)$ is approximated by

$$\overset{(e)}{y}(x) = \overset{(e)}{y_1}\overset{(e)}{N_1}(x) + \overset{(e)}{y_1'}\overset{(e)}{N_2}(x) + \overset{(e)}{y_2}\overset{(e)}{N_3}(x) + \overset{(e)}{y_2'}\overset{(e)}{N_4}(x) \qquad (4.112)$$

where $\overset{(e)}{y_1}, \overset{(e)}{y_1'}, \overset{(e)}{y_2}$, and $\overset{(e)}{y_2'}$ are the values of $y(x)$ and $y'(x)$ at nodes 1 and 2 of element (e). Then Equation (4.112) may be readily put into the form of Equation (4.37). That is,

$$y(x) = \sum_{e=1}^{n} \overset{(e)}{y}(x) = \sum_{e=1}^{n} \{\overset{(e)}{y}\}^T\{\overset{(e)}{N(x)}\} = \sum_{e=1}^{n} \{\overset{(e)}{N(x)}\}^T\{\overset{(e)}{y}\}$$

$$(4.113)$$

where $\{\overset{(e)}{y}\}$ and $\{\overset{(e)}{N(x)}\}$ are now the arrays

$$\{\overset{(e)}{y}\} = \begin{Bmatrix} \overset{(e)}{y_1} \\ \overset{(e)}{y_1'} \\ \overset{(e)}{y_2} \\ \overset{(e)}{y_2'} \end{Bmatrix} \qquad \{\overset{(e)}{N(x)}\} = \begin{Bmatrix} \overset{(e)}{N_1}(x) \\ \overset{(e)}{N_2}(x) \\ \overset{(e)}{N_3}(x) \\ \overset{(e)}{N_4}(x) \end{Bmatrix} \qquad (4.114)$$

4.11 EXAMPLE APPLICATION WITH HERMITIAN INTERPOLATES

To see how the Hermitian interpolates may be used in an application, consider again the example boundary value problem of Sections 4.1 and 4.8. Recall that the solution to the problem

$$\frac{d^2y}{dx^2} - y = f(x) \qquad (4.115)$$

where

$$y(a) = c \quad \text{and} \quad y(b) = d \qquad (4.116)$$

may be obtained by finding $y(x)$ which minimizes the integral

$$I = \int_a^b [y^2 + (y')^2 + 2yf]\ dx \tag{4.117}$$

Hence, by dividing the interval $[a, b]$ into n elements and by representing $y(x)$ over each element as in Equation (4.112) and (4.113), I may be written [see Equation (4.67)]

$$I = \int_a^b \left[\sum_{e=1}^{n} \{\overset{(e)}{y}\}^T \{\overset{(e)}{N}\} \{\overset{(e)}{N}\}^T \{\overset{(e)}{y}\} + \{\overset{(e)}{y}\}^T \{\overset{(e)}{N'}\}^T \{\overset{(e)}{N'}\} \{\overset{(e)}{y}\} \right.$$
$$\left. + 2\{\overset{(e)}{y}\}^T \{\overset{(e)}{N}\} \{\overset{(e)}{N}\}^T \{\overset{(e)}{f}\} \right] dx \tag{4.118}$$

or as

$$I = \sum_{e=1}^{n} \overset{(e)}{I} \tag{4.119}$$

where

$$\overset{(e)}{I} = \{\overset{(e)}{y}\}^T [\overset{(e)}{k}] \{\overset{(e)}{y}\} + 2\{\overset{(e)}{y}\}^T \{\overset{(e)}{g}\} \tag{4.120}$$

where, as before $[\overset{(e)}{k}]$ and $\{\overset{(e)}{g}\}$ are defined as [see Equations (4.70) and (4.71)]

$$[\overset{(e)}{k}] = \int_a^b [\{\overset{(e)}{N}\}\{\overset{(e)}{N}\}^T + \{\overset{(e)}{N'}\}\{\overset{(e)}{N'}\}^T]\ dx \tag{4.121}$$

and

$$\{\overset{(e)}{g}\} = \left(\int_a^b [\{\overset{(e)}{N}\} \{\overset{(e)}{N}\}^T]\ dx \right) \{\overset{(e)}{f}\} \tag{4.122}$$

Since in Equations (4.113) and (4.114) $\{\overset{(e)}{N}\}$ is an array with 4 entries, the matrix of Equation (4.121) is a 4×4 matrix (as opposed to the 2×2 matrix of Section 4.8). Specifically, $[\overset{(e)}{k}]$ and $\{\overset{(e)}{g}\}$ become

$$
\overset{(e)}{[k]} = \int_a^b \begin{bmatrix}
\overset{(e)}{N_1}\overset{(e)}{N_1}+\overset{(e)}{N'_1}\overset{(e)}{N'_1} & \overset{(e)}{N_1}\overset{(e)}{N_2}+\overset{(e)}{N'_1}\overset{(e)}{N'_2} & \overset{(e)}{N_1}\overset{(e)}{N_3}+\overset{(e)}{N'_1}\overset{(e)}{N'_3} & \overset{(e)}{N_1}\overset{(e)}{N_4}+\overset{(e)}{N'_1}\overset{(e)}{N'_4} \\[6pt]
\overset{(e)}{N_2}\overset{(e)}{N_1}+\overset{(e)}{N'_2}\overset{(e)}{N'_1} & \overset{(e)}{N_2}\overset{(e)}{N_2}+\overset{(e)}{N'_2}\overset{(e)}{N'_2} & \overset{(e)}{N_2}\overset{(e)}{N_3}+\overset{(e)}{N'_2}\overset{(e)}{N'_3} & \overset{(e)}{N_2}\overset{(e)}{N_4}+\overset{(e)}{N'_2}\overset{(e)}{N'_4} \\[6pt]
\overset{(e)}{N_3}\overset{(e)}{N_1}+\overset{(e)}{N'_3}\overset{(e)}{N'_1} & \overset{(e)}{N_3}\overset{(e)}{N_2}+\overset{(e)}{N'_3}\overset{(e)}{N'_2} & \overset{(e)}{N_3}\overset{(e)}{N_3}+\overset{(e)}{N'_3}\overset{(e)}{N'_3} & \overset{(e)}{N_3}\overset{(e)}{N_4}+\overset{(e)}{N'_3}\overset{(e)}{N'_4} \\[6pt]
\overset{(e)}{N_4}\overset{(e)}{N_1}+\overset{(e)}{N'_4}\overset{(e)}{N'_1} & \overset{(e)}{N_4}\overset{(e)}{N_2}+\overset{(e)}{N'_4}\overset{(e)}{N'_2} & \overset{(e)}{N_4}\overset{(e)}{N_3}+\overset{(e)}{N'_4}\overset{(e)}{N'_3} & \overset{(e)}{N_4}\overset{(e)}{N_4}+\overset{(e)}{N'_4}\overset{(e)}{N'_4}
\end{bmatrix} dx
$$

$$(4.123)$$

and

$$
\overset{(e)}{\{g\}} = \left(\int_a^b \begin{bmatrix}
\overset{(e)}{N_1}\overset{(e)}{N_1} & \overset{(e)}{N_1}\overset{(e)}{N_2} & \overset{(e)}{N_1}\overset{(e)}{N_3} & \overset{(e)}{N_1}\overset{(e)}{N_4} \\[6pt]
\overset{(e)}{N_2}\overset{(e)}{N_1} & \overset{(e)}{N_2}\overset{(e)}{N_2} & \overset{(e)}{N_2}\overset{(e)}{N_3} & \overset{(e)}{N_2}\overset{(e)}{N_4} \\[6pt]
\overset{(e)}{N_3}\overset{(e)}{N_1} & \overset{(e)}{N_3}\overset{(e)}{N_2} & \overset{(e)}{N_3}\overset{(e)}{N_3} & \overset{(e)}{N_3}\overset{(e)}{N_4} \\[6pt]
\overset{(e)}{N_4}\overset{(e)}{N_1} & \overset{(e)}{N_4}\overset{(e)}{N_2} & \overset{(e)}{N_4}\overset{(e)}{N_3} & \overset{(e)}{N_4}\overset{(e)}{N_4}
\end{bmatrix} dx \right) \overset{(e)}{\{f\}} \quad (4.124)
$$

To explicitly determine $\overset{(e)}{[k]}$ and $\overset{(e)}{\{g\}}$ it is thus necessary to evaluate a number of integrals of the element Hermitian interpolates and their derivatives. Because of the complexity of the interpolates as defined in Equation (4.109), the evaluation of these integrals can be tedious and detailed. Consider, for example, $\int_a^b \overset{(e)}{N_1}\overset{(e)}{N_1}\,dx$. This integral becomes

$$
\int_a^b \overset{(e)}{N_1}\overset{(e)}{N_1}\,dx = \int_{\overset{(e)}{x_1}}^{\overset{(e)}{x_2}} (x-\overset{(e)}{x_2})^4
$$

$$(4.125)$$

$$
\times\ [(\overset{(e)}{x_1}-\overset{(e)}{x_2})+2(\overset{(e)}{x_1}-x)]^2/(\overset{(e)}{x_1}-\overset{(e)}{x_2})^6\,dx
$$

The integrand is a sixth order polynomial in x with lengthy coefficients. However, this integral, as well as the others in Equations (4.123) and (4.124) may be simplified considerably by making the substitutions

$$
x = x_1 + \xi \quad \text{and} \quad \ell = \overset{(e)}{x_2} - \overset{(e)}{x_1} \tag{4.126}
$$

Then $\overset{(e)}{N_1}(x)$ becomes $\overset{(e)}{N_1}(\xi)$. That is,

$$\overset{(e)}{N_1}(x) \rightarrow \overset{(e)}{N_1}(\xi) = (\xi - \ell)^2(\ell + 2\xi)/\ell^3 \tag{4.127}$$

and the integral of Equation (4.125) becomes

$$\int_a^b N_1 N_1 \, dx = \int_0^\ell N_1 N_1 \, d\xi = \int_0^\ell [(\xi - \ell)^4(\ell + 2\xi)^2/\ell^6] \, d\xi$$

$$= 13\ell/35 \tag{4.128}$$

Similarly, in terms of ξ and ℓ and the other element Hermitian interpolating functions are

$$\overset{(e)}{N_2}(\xi) = \xi(\xi - \ell)^2/\ell^2$$

$$\overset{(e)}{N_3}(\xi) = \xi^2(3\ell - 2\xi)/\ell^3$$

$$\overset{(e)}{N_4}(\xi) = \xi^2(\xi - \ell)/\ell^2$$

$$\overset{(e)}{N_1'}(\xi) = (6\xi/\ell^2)[(\xi/\ell) - 1]$$

$$\overset{(e)}{N_2'}(\xi) = 1 - 4(\xi/\ell) + 3(\xi/\ell)^2$$

$$\overset{(e)}{N_3'}(\xi) = (6\xi/\ell^2)[1 - (\xi/\ell)] \tag{4.129}$$

$$\overset{(e)}{N_4'}(\xi) = (\xi/\ell)[(3\xi/\ell) - 2]$$

$$\overset{(e)}{N_1''}(\xi) = (6/\ell^2)[(2\xi/\ell) - 1]$$

$$\overset{(e)}{N_2''}(\xi) = (2/\ell)[(3\xi/\ell) - 2]$$

$$\overset{(e)}{N_3''}(\xi) = (6/\ell^2)[1 - (2\xi/\ell)]$$

$$\overset{(e)}{N_4''}(\xi) + (2/\ell)[(3\xi/\ell) - 1]$$

The various integrals of these functions are then found to be

$$\int_a^b N_1^{(e)} N_2^{(e)} \, dx = \int_0^\ell N_1^{(e)} N_2^{(e)} \, d\xi = 11\ell^2/210$$

$$\int_a^b N_1^{(e)} N_3^{(e)} \, dx = \int_0^\ell N_1^{(e)} N_3^{(e)} \, d\xi = 9\ell/70$$

$$\int_a^b N_1^{(e)} N_4^{(e)} \, dx = \int_0^\ell N_1^{(e)} N_4^{(e)} \, d\xi = -13\ell^2/420$$

$$\int_a^b N_2^{(e)} N_2^{(e)} \, dx = \int_0^\ell N_2^{(e)} N_2^{(e)} \, d\xi = \ell^3/105$$

$$\int_a^b N_2^{(e)} N_3^{(e)} \, dx = \int_0^\ell N_2^{(e)} N_3^{(e)} \, d\xi = 13\ell^2/420$$

$$\int_a^b N_2^{(e)} N_4^{(e)} \, dx = \int_0^\ell N_2^{(e)} N_4^{(e)} \, d\xi = -\ell^3/140$$

$$\int_a^b N_3^{(e)} N_3^{(e)} \, dx = \int_0^\ell N_3^{(e)} N_3^{(e)} \, d\xi = 13\ell/35$$

$$\int_a^b N_3^{(e)} N_4^{(e)} \, dx = \int_0^\ell N_3^{(e)} N_4^{(e)} \, d\xi = -11\ell^2/210$$

$$\int_a^b N_4^{(e)} N_4^{(e)} \, dx = \int_0^\ell N_4^{(e)} N_4^{(e)} \, d\xi = \ell^3/105$$

$$\int_a^b N_1'^{(e)} N_1'^{(e)} \, dx = \int_0^\ell N_1'^{(e)} N_1'^{(e)} \, d\xi = 6/5\ell$$

$$\int_a^b N_1'^{(e)} N_2'^{(e)} \, dx = \int_0^\ell N_1'^{(e)} N_2'^{(e)} \, d\xi = 1/10$$

$$\int_a^b N_1'^{(e)} N_3'^{(e)} \, dx = \int_0^\ell N_1'^{(e)} N_3'^{(e)} \, d\xi = -6/5\ell$$

$$\int_a^b N_1'^{(e)} N_4'^{(e)} \, dx = \int_0^\ell N_1'^{(e)} N_4'^{(e)} \, d\xi = 1/10$$

$$\int_a^b \overset{(e)}{N_2'} \overset{(e)}{N_2'} \, dx = \int_0^\ell \overset{(e)}{N_2'} \overset{(e)}{N_2'} \, d\xi = 2\ell/15$$

$$\int_a^b \overset{(e)}{N_2'} \overset{(e)}{N_3'} \, dx = \int_0^\ell \overset{(e)}{N_2'} \overset{(e)}{N_3'} \, d\xi = -1/10$$

$$\int_a^b \overset{(e)}{N_2'} \overset{(e)}{N_4'} \, dx = \int_0^\ell \overset{(e)}{N_2'} \overset{(e)}{N_4'} \, d\xi = -\ell/30$$

$$\int_a^b \overset{(e)}{N_3'} \overset{(e)}{N_3'} \, dx = \int_0^\ell \overset{(e)}{N_3'} \overset{(e)}{N_3'} \, d\xi = 6/5\ell$$

$$\int_a^b \overset{(e)}{N_3'} \overset{(e)}{N_4'} \, dx = \int_0^\ell \overset{(e)}{N_3'} \overset{(e)}{N_4'} \, d\xi = -1/10$$

$$\int_a^b \overset{(e)}{N_4'} \overset{(e)}{N_4'} \, dx = \int_0^\ell \overset{(e)}{N_4'} \overset{(e)}{N_4'} \, d\xi = 2\ell/15$$

$$\int_a^b \overset{(e)}{N_1''} \overset{(e)}{N_1''} \, dx = \int_0^\ell \overset{(e)}{N_1''} \overset{(e)}{N_1''} \, d\xi = 12/\ell^3$$

$$\int_a^b \overset{(e)}{N_1''} \overset{(e)}{N_2''} \, dx = \int_0^\ell \overset{(e)}{N_1''} \overset{(e)}{N_2''} \, d\xi = 6/\ell^2$$

$$\int_a^b \overset{(e)}{N_1''} \overset{(e)}{N_3''} \, dx = \int_0^\ell \overset{(e)}{N_1''} \overset{(e)}{N_3''} \, d\xi = -12/\ell^3$$

$$\int_a^b \overset{(e)}{N_1''} \overset{(e)}{N_4''} \, dx = \int_0^\ell \overset{(e)}{N_1''} \overset{(e)}{N_4''} \, d\xi = 6/\ell^2 \qquad (4.130)$$

$$\int_a^b \overset{(e)}{N_2''} \overset{(e)}{N_2''} \, dx = \int_0^\ell \overset{(e)}{N_2''} \overset{(e)}{N_2''} \, d\xi = 4/\ell$$

$$\int_a^b \overset{(e)}{N_2''} \overset{(e)}{N_3''} \, dx = \int_0^\ell \overset{(e)}{N_2''} \overset{(e)}{N_3''} \, d\xi = -6/\ell$$

$$\int_a^b \overset{(e)}{N_2''} \overset{(e)}{N_4''} \, dx = \int_0^\ell \overset{(e)}{N_2''} \overset{(e)}{N_4''} \, d\xi = 2/\ell$$

$$\int_a^b \overset{(e)}{N_3''} \overset{(e)}{N_3''} \, dx = \int_0^\ell \overset{(e)}{N_3''} \overset{(e)}{N_3''} \, d\xi = 12/\ell^3$$

$$\int_a^b \overset{(e)}{N_3''} \overset{(e)}{N_4''} \, dx = \int_0^\ell \overset{(e)}{N_3''} \overset{(e)}{N_4''} \, d\xi = -6/\ell^2$$

$$\int_a^b \overset{(e)}{N_4''} \overset{(e)}{N_4''} \, dx = \int_0^\ell \overset{(e)}{N_4''} \overset{(e)}{N_4''} \, d\xi = 4/\ell$$

(The integrals of the second derivatives of the element Hermitian interpolates are included in this listing since they are useful in fourth order differential equations as in beam theory. This is explored in the following section.)

By using Equations (4.130), the element stiffness matrix $[\overset{(e)}{k}]$ of Equation (4.123) becomes

$$
[\overset{(e)}{k}] =
\begin{bmatrix}
13\ell/35 & 11\ell^2/210 & 9\ell/70 & -13\ell^2/420 \\
11\ell^2/210 & \ell^3/105 & 13\ell^2/420 & -\ell^3/140 \\
9\ell/70 & 13\ell^2/420 & 13\ell/35 & -11\ell^2/210 \\
-13\ell^2/420 & -\ell^3/140 & -11\ell^2/210 & \ell^3/105
\end{bmatrix}
$$

$$
+
\begin{bmatrix}
6/5\ell & 1/10 & -6/5\ell & 1/10 \\
1/10 & 2\ell/15 & -1/10 & -\ell/30 \\
-6/5\ell & -1/10 & 6/5\ell & -1/10 \\
1/10 & -\ell/30 & -1/10 & 2\ell/15
\end{bmatrix}
\qquad (4.131)
$$

Similarly, $\{\overset{(e)}{g}\}$ becomes

$$
\{\overset{(e)}{g}\} =
\begin{bmatrix}
13\ell/35 & 11\ell^2/210 & 9\ell/70 & -13\ell^2/420 \\
11\ell^2/210 & \ell^3/105 & 13\ell^2/420 & -\ell^3/140 \\
9\ell/70 & 13\ell^2/420 & 13\ell/35 & -11\ell^2/210 \\
-13\ell^2/420 & -\ell^3/140 & -11\ell^2/210 & \ell^3/105
\end{bmatrix}
\{\overset{(e)}{f}\}
$$

$$(4.132)$$

Note that the array $\left\{ \overset{(e)}{f} \right\}$ in Equation (4.132) includes derivatives of the given function at the nodes of the element.

The element stiffness matrix and the forcing array may be expanded into global dimensions as in Section 4.7 by using the incidence matrices. That is,

$$\overset{(e)}{[K]} = \overset{(e)}{[\Lambda]} \overset{(e)}{[k]} \overset{(e)}{[\Lambda]}^T \tag{4.133}$$

and

$$\left\{ \overset{(e)}{G} \right\} = \overset{(e)}{[\Lambda]} \left\{ \overset{(e)}{g} \right\} \tag{4.134}$$

In terms of these arrays, the integral I of Equation (4.119) becomes

$$I = \{Y\}^T [K]\{Y\} + 2 \{Y\}^T \{G\} \tag{4.135}$$

where $[K]$ and $\{G\}$ are

$$[K] = \sum_{e=1}^{n} \overset{(e)}{[K]} \tag{4.136}$$

and

$$\{G\} = \sum_{e=1}^{n} \left\{ \overset{(e)}{G} \right\} \tag{4.137}$$

Choosing the Y_i of Y in Equation (4.135) such that I is a minimum (that is, setting the derivative of I with respect to the Y_i equal to zero) leads to the now familiar expression

$$[K]\{Y\} = -\{G\} \tag{4.138}$$

To illustrate the use of the above expressions, let us consider the same boundary value problem [see Equations (4.115) and (4.116)] that we studied in Section 4.8 [see Equations (4.87), (4.88), and (4.89)]. That is, let $[a, b]$ be $[0, 4]$, let $f(x)$ be f_0, a constant and let c and d be $1 - f_0$ and $e^4 - f_0$, respectively. Further, let the interval $[0, 4]$ be divided into two elements each of length 2.

With 2 elements, the incidence matrices become

$$
\overset{(1)}{[\Lambda]} =
\begin{bmatrix}
1 & 0 & 0 & 0 \\
0 & 1 & 0 & 0 \\
0 & 0 & 1 & 0 \\
0 & 0 & 0 & 1 \\
0 & 0 & 0 & 0 \\
0 & 0 & 0 & 0
\end{bmatrix}
\quad \text{and} \quad
\overset{(2)}{[\Lambda]} =
\begin{bmatrix}
0 & 0 & 0 & 0 \\
0 & 0 & 0 & 0 \\
1 & 0 & 0 & 0 \\
0 & 1 & 0 & 0 \\
0 & 0 & 1 & 0 \\
0 & 0 & 0 & 1
\end{bmatrix}
\tag{4.139}
$$

Then $[K]$ becomes (see Problem 4.18)

$$
[K] =
\begin{bmatrix}
47/35 & 13/42 & -12/35 & -1/42 & 0 & 0 \\
13/42 & 36/105 & 1/42 & -13/105 & 0 & 0 \\
-12/35 & 1/42 & 94/35 & 0 & -12/35 & -1/42 \\
-1/42 & -13/105 & 0 & 24/35 & 1/42 & -13/105 \\
0 & 0 & -12/35 & 1/42 & 47/35 & -13/42 \\
0 & 0 & -1/42 & -13/105 & -13/42 & 36/105
\end{bmatrix}
\tag{4.140}
$$

Similarly, $\{G\}$ becomes (see Problem 4.19)

$$
\{G\} =
\begin{bmatrix}
26/35 & 22/105 & 9/35 & -13/105 & 0 & 0 \\
22/105 & 8/105 & 13/105 & -2/35 & 0 & 0 \\
9/35 & 13/105 & 52/35 & 0 & 9/35 & -13/105 \\
-13/105 & -2/35 & 0 & 16/105 & 13/105 & -2/35 \\
0 & 0 & 9/35 & 13/105 & 26/35 & -22/105 \\
0 & 0 & -13/105 & -2/35 & -22/105 & 8/105
\end{bmatrix}
\begin{Bmatrix}
f_0 \\ 0 \\ f_0 \\ 0 \\ f_0 \\ 0
\end{Bmatrix}
=
\begin{Bmatrix}
f_0 \\ f_0/3 \\ 2f_0 \\ 0 \\ f_0 \\ -f_0/3
\end{Bmatrix}
\tag{4.141}
$$

Hence, the governing equation (4.138) becomes

$$
\begin{bmatrix}
47/35 & 13/42 & -12/35 & -1/42 & 0 & 0 \\
13/42 & 12/35 & 1/42 & -13/105 & 0 & 0 \\
-12/35 & 1/42 & 94/35 & 0 & -12/35 & -1/42 \\
-1/42 & -13/105 & 0 & 24/35 & 1/42 & -13/105 \\
0 & 0 & -12/35 & 1/42 & 47/35 & -13/42 \\
0 & 0 & -1/42 & -13/105 & -13/42 & 12/35
\end{bmatrix}
\begin{Bmatrix}
Y_0 \\ Y_0' \\ Y_1 \\ Y_1' \\ Y_2 \\ Y_2'
\end{Bmatrix}
= -f_0
\begin{Bmatrix}
1 \\ 1/3 \\ 2 \\ 0 \\ 1 \\ -1/3
\end{Bmatrix}
$$

$$(4.142)$$

As in Section 4.8, we may express the solution to Equation (4.142) in the form

$$\{Y\} = \{Y_H\} + \{Y_P\} \tag{4.143}$$

where $\{Y_H\}$ is a solution to

$$
\begin{bmatrix}
47/35 & 13/42 & -12/35 & -1/42 & 0 & 0 \\
13/42 & 12/35 & 1/42 & -13/105 & 0 & 0 \\
-12/35 & 1/42 & 94/35 & 0 & -12/35 & -1/42 \\
-1/42 & -13/105 & 0 & 24/35 & 1/42 & -13/105 \\
0 & 0 & -12/35 & 1/42 & 47/35 & -13/42 \\
0 & 0 & -1/42 & -13/105 & -13/42 & 12/35
\end{bmatrix}
\begin{Bmatrix}
Y_{H0} \\ Y_{H0}' \\ Y_{H1} \\ Y_{H1}' \\ Y_{H2} \\ Y_{H2}'
\end{Bmatrix}
= 0
$$

$$(4.144)$$

Recall, however, that the values of y are specified at the end-points of the interval. Then the first and fifth entries of $\{Y_H\}$ (that is, Y_{H0} and Y_{H2}) are determined from these values. Hence, from the matrix equation (4.144) we obtain (omitting the first and fifth equations)

$$(12/35) \, Y'_{H0} + (1/42) \, Y_{H1} - (13/105) \, Y'_{H1}$$
$$+ 0 \, Y'_{H2} = - (13/42) \, Y_{H0}$$

$$(1/42) \, Y'_{H0} + (94/35) \, Y_{H1} + 0 \, Y'_{H1} - (1/42) \, Y'_{H2}$$
$$= (12/35) \, Y_{H0} + (12/35) \, Y_{H2}$$

$$- (13/105) \, Y'_{H0} + 0 \, Y_{H1} + (24/35) \, Y'_{H1} - (13/105) \, Y'_{H2}$$
$$= (1/42) \, Y_{H0} - (1/42) \, Y_{H2}$$

$$0 \, Y'_{H0} + (1/42) \, Y_{H1} + (13/105) \, Y'_{H1} - (13/35) \, Y'_{H2}$$
$$= - (13/42) \, Y_{H2}$$

(4.145)

Since $\{Y_P\}$ is any particular solution of Equation (4.142), it is easily verified that such a solution is

$$\{Y_P\} = \begin{Bmatrix} f_0 \\ 0 \\ f_0 \\ 0 \\ f_0 \\ 0 \end{Bmatrix}$$

(4.146)

Hence, by using Equation (4.143) to satisfy the boundary conditions Y_{H0} and Y_{H2} are found to be

$$Y_{H0} = 1 \quad \text{and} \quad Y_{H2} = e^4 = 54.598$$

(4.147)

Then, by solving Equations (4.145) for Y'_{H0}, Y_{H1}, Y'_{H1}, and Y'_{H2}, we obtain

$$Y'_{H0} = 1.428$$

$$Y_{H1} = 7.552$$

$$Y'_{H1} = 7.906$$
(4.148)

$$Y'_{H2} = 52.558$$

Finally, from Equation (4.143), the solution for $\{Y\}$ becomes

$$Y_0 = Y_{H0} + Y_{P0} = 1 - f_0$$

$$Y'_0 = Y'_{H0} + Y'_{P0} = 1.428$$

$$Y_1 = Y_{H1} + Y_{P1} = 7.552 - f_0$$

$$Y'_1 = Y'_{H1} + Y'_{P1} = 7.906$$
(4.149)

$$Y_2 = Y_{H2} + P_{P2} = 54.598 - f_0$$

$$Y'_2 = Y'_{H2} + P'_{P2} = 52.668$$

From Equation (4.89), Section 4.8, we find the exact solution to be

$$Y_0 = 1 - f_0$$

$$Y'_0 = 1.0$$

$$Y_1 = 7.389 - f_0$$

$$Y'_1 = 7.389$$
(4.150)

$$Y_2 = 54.598 - f_0$$

$$Y'_2 = 54.598$$

For only 2 elements this is a reasonably accurate result. However, by comparing the procedure using the Hermitian interpolates with the solution of Section 4.8 which uses the Lagrange interpolates, we see that the presence of the derivatives in the $\{Y\}$ array makes the use of the Hermitian interpolates more cumbersome. Indeed, we obtained better accuracy with less effort with the Lagrange interpolates. Hence, unless we are concerned with the values of the derivatives of the solution function, we find the Lagrange interpolates to be preferable. This conclusion is reversed, however, if we consider 4th order boundary value problems as are encountered, for example, in beam theory. Noting that the solution of $d^4y/dx^4 = 0$ is a cubic polynomial, we would expect to obtain an excellent approximation with the Hermitian interpolates. Moreover, the Hermitian interpolates are "natural" functions for satisfying boundary conditions for 4th order equations, especially where the derivatives are specified in the boundary conditions. The development of these ideas are explored in the following section.

4.12   USE OF HERMITIAN INTERPOLATES
IN BEAM MODELING

In this section, we will further explore the use of the Hermitian interpolation functions. We will use thin beam theory to illustrate the use of the functions. To simplify our analysis we will neglect axial loading on the beam elements. That is, we will restrict our attention to bending phenomena.

Recall that the governing differential equation for the vertical displacement of a beam element is

$$EI\, d^4y/dx^4 = p(x) \tag{4.151}$$

where $p(x)$ is the loading on the beam element and is taken as positive in the positive Y direction (see Figure 4.15).

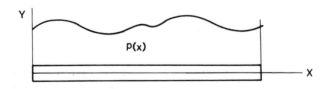

Figure 4.15   Beam Element and Loading Function

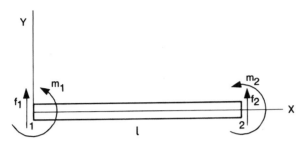

Figure 4.16   Beam Element and End Loadings

Consider a beam element of length ℓ as in Chapter 3. Suppose the forces and moments at the ends of the beam element are equivalent to f_1, m_1, f_2, and m_2 as shown in Figure 4.16. Then the governing differential equation for the beam element is Equation (4.151) together with the boundary conditions

$$EI\, d^2 y/dx^2 = -m_1 \quad \text{and} \quad EI\, d^3 y/dx^3 = f_1 \quad \text{at} \quad x = 0 \quad (4.152)$$

and

$$EI\, d^2 y/dx^2 = m_2 \quad \text{and} \quad EI\, d^3 y/dx^3 = -f_2 \quad \text{at} \quad x = \ell \quad (4.153)$$

Recall that in Chapter 1 we found that an equivalent formulation of this boundary value problem is: Find $y(x)$ such that the integral J is a minimum where J is given by

$$J = (\tfrac{1}{2}) \int_0^\ell [EI(y'')^2 - 2yp]\, dx - f_1 y(0) - m_1 y'(0) - f_2 y(\ell)$$
$$- m_2 y'(\ell) \qquad (4.154)$$

where the notation is the same as that used previously. We can solve this extremum problem by following the same procedures developed in the earlier sections of this chapter. That is, over the beam element, let $y(x)$ be written in the form

$$y(x) = \{\overset{(e)}{y}\}^T \{\overset{(e)}{N}(x)\} \qquad (4.155)$$

where $\{\overset{(e)}{y}\}^T$ is given by

$$\{\overset{(e)}{y}\}^T = [u_1, \theta_1, u_2, \theta_2] \tag{4.156}$$

where u_1, θ_1, u_2, and θ_2 are the element displacements and rotations at the respective ends. $\{\overset{(e)}{N(x)}\}$ is the element Hermitian interpolate array given by

$$\{\overset{(e)}{N(x)}\} = \begin{bmatrix} \overset{(e)}{N_1}(x) \\ \overset{(e)}{N_2}(x) \\ \overset{(e)}{N_3}(x) \\ \overset{(e)}{N_4}(x) \end{bmatrix} \tag{4.157}$$

where $\overset{(e)}{N_1}$, $\overset{(e)}{N_2}$, $\overset{(e)}{N_3}$, and $\overset{(e)}{N_4}$ are the Hermitian interpolates given by Equation (4.109). [Regarding notation, in this case there is no summation needed in Equation (4.155) since the element itself is the entire beam.] By substituting from Equation (4.155) into Equation (4.154) we obtain for J

$$J = (EI/2) \{\overset{(e)}{y}\}^T \left[\int_0^{\ell} \overset{(e)}{N''}{}^T \overset{(e)}{N''} dx \right] \{\overset{(e)}{y}\}$$

$$- \{\overset{(e)}{y}\}^T \int_0^{\ell} \{\overset{(e)}{N}\} \, p(x) \, dx - \{\overset{(e)}{y}\}^T \{\overset{(e)}{f}\} \tag{4.158}$$

where $\{\overset{(e)}{f}\}$ is the boundary loading array

$$\{\overset{(e)}{f}\} = \begin{Bmatrix} f_1 \\ m_1 \\ f_2 \\ m_2 \end{Bmatrix} \tag{4.159}$$

Equation (4.158) may be rewritten in the form

$$J = \tfrac{1}{2} \{\overset{(e)}{y}\}^T [\overset{(e)}{k}] \{\overset{(e)}{y}\} - \{\overset{(e)}{y}\}^T \{\overset{(e)}{p}\} - \{\overset{(e)}{y}\}^T \{\overset{(e)}{f}\} \tag{4.160}$$

where $\{\overset{(e)}{p}\}$ is defined as

$$\{\overset{(e)}{p}\} = \int_0^\ell \{\overset{(e)}{N}\} \, p(x) \, dx \tag{4.161}$$

and where the element stiffness matrix $[\overset{(e)}{k}]$ is found to be

$$[\overset{(e)}{k}] = EI \int_0^\ell \{\overset{(e)}{N''}\}^T \{\overset{(e)}{N''}\} \, dx = (EI/\ell^3) \begin{bmatrix} 12 & 6\ell & -12 & 6\ell \\ 6\ell & 4\ell^2 & -6\ell & 2\ell^2 \\ -12 & -6\ell & 12 & -6\ell \\ 6\ell & 2\ell^2 & -6\ell & 4\ell^2 \end{bmatrix} \tag{4.162}$$

The result is obtained directly from Equation (4.130) of the previous section.

It is interesting (and reassuring) to observe that the element stiffness array of Equation (4.162) is exactly the same as the element stiffness array of Equation (3.34) of Chapter 3 when the axial loading is neglected.

The solution of the boundary value problem may now be obtained in the usual manner by setting the derivatives of I in Equation (4.160) with respect to the components of $\{\overset{(e)}{y}\}$ equal to zero. This leads immediately to the relations

$$[\overset{(e)}{k}]\{\overset{(e)}{y}\} = \{\overset{(e)}{f}\} + \{\overset{(e)}{p}\} \tag{4.163}$$

A comparison of this result with Equation (3.79) of Chapter 3 shows that $\{\overset{(e)}{p}\}$ is simply the "fixed-end" load array. Hence, Equation (4.161) provides an especially convenient algorithm for determining the fixed-end force array. To see this, imagine a beam with fixed ends and consider the integral of Equation (4.161). By using Equation (4.151), this integral may be rewritten

$$\begin{Bmatrix} (e) \\ p \end{Bmatrix} = \int_0^\ell \begin{Bmatrix} (e) \\ N \end{Bmatrix} p(x)\,dx = EI \int_0^\ell \begin{Bmatrix} (e) \\ N \end{Bmatrix} (d^4y/dx^4)\,dx \qquad (4.164)$$

Examine the first term in this array. That is, consider

$$p_1 = \int_0^\ell \overset{(e)}{N_1} p(x)\,dx = EI \int_0^\ell \overset{(e)}{N_1} (d^4y/dx^4)\,dx \qquad (4.165)$$

Recall from Equation (4.111) that $\overset{(e)}{N_1}(0) = 1$, while $\overset{(e)}{N_1}(\ell) = \overset{(e)}{N_1'}(0) =$ $\overset{(e)}{N_1'}(\ell) = 0$. Hence, by integrating Equation (4.165) by parts, we obtain

$$p_1 = EI\,\overset{(e)}{N_1}(d^3y/dx^3)\,\Big|_0^\ell - EI \int_0^\ell \overset{(e)}{N_1'}(d^3y/dx^3)\,dx$$

$$= 0 - EI(d^3y/dx^3)\big|_{x=0} - EI\,\cancel{\overset{(e)}{N_1'}}\,d^2y/dx^2\,\Big|_0^\ell$$

$$+ EI \int_0^\ell \overset{(e)}{N''}(d^2y/dx^2)\,dx$$

$$= - EI(d^3y/dx^3)\,\big|_{x=0} + EI\,\overset{(e)}{N'''}(d\cancel{y}/dx)\,\Big|_0^\ell$$

$$- EI \int_0^\ell \overset{(e)}{N'''}(dy/dx)\,dx$$

$$= - EI(d^3y/dx^3)\,\big|_{x=0} - EI\,\overset{(e)}{N'''}\cancel{y}\,\Big|_0^\ell$$

$$+ EI \int_0^\ell \cancel{\overset{(e)}{N''''}}\, y\,dx$$

or

$$\overset{(e)}{p_1} = - EI(d^3y/dx^3)\,\big|_{x=0} = -f_1 \qquad (4.166)$$

[The second equality in Equation (4.166) follows immediately from Equation (4.152). Note also from Equation (4.109) that since $\overset{(e)}{N_1}(x)$ is a cubic in x, $N_1''''(x) = 0$.]

By similar analyses, it is easily shown that

$$\overset{(e)}{p_2} = EI(d^2y/dx^2)\,|_{x=0} = -m_1$$

$$\overset{(e)}{p_3} = EI(d^3y/dx^3)\,|_{x=\ell} = -f_2 \qquad\qquad (4.167)$$

$$\overset{(e)}{p_4} = -EI(d^2y/dx^2)\,|_{x=\ell} = -m_2$$

Hence, $\{\overset{(e)}{p}\}$ may be written

$$\{\overset{(e)}{p}\} = -\begin{Bmatrix} f_1 \\ m_1 \\ f_2 \\ m_2 \end{Bmatrix} = -\{\overset{(e)}{r}\} \qquad\qquad (4.168)$$

where $\{\overset{(e)}{r}\}$ is the support reaction array. However, by recalling from Chapter 3 that the "fixed-end" load array is the negative of the support reaction array. Equation (4.168) shows that, as stated above, the "fixed-end" load array is $\{\overset{(e)}{p}\}$.

4.13 APPLICATIONS IN BEAM MODELING

As a first application of the foregoing results, let us determine the "fixed-end" load array for the loading shown in Figure 4.17.

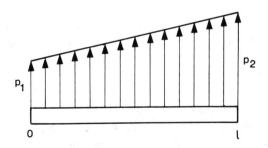

Figure 4.17 Linearly Varying Load Array

That is, let $p(x)$ be expressed as

$$p(x) = (p_2 - p_1)(x/\ell) + p_1 \tag{4.169}$$

Then, by using Equation (4.164) together with the definitions of the Hermitian interpolation functions of Equations (4.127) and (4.129), the "fixed-end" load array p becomes

$$\{p\} = \begin{bmatrix} (7/20)p_1\ell + (3/20)p_2\ell \\ (1/20)p_1\ell^2 + (1/30)p_2\ell^2 \\ (3/20)p_1\ell + (7/20)p_2\ell \\ -(1/30)p_1\ell^2 - (1/20)p_2\ell^2 \end{bmatrix} \tag{4.170}$$

For the special case when $p_1 = p_2 = p_0$, that is with uniform loading, $\{p\}$ becomes

$$\{p\} = \begin{bmatrix} (1/2)p_0\ell \\ (1/12)p_0\ell^2 \\ (1/2)p_0\ell \\ -(1/12)p_0\ell^2 \end{bmatrix} \tag{4.171}$$

These results are identical to those obtained earlier in Chapter 3 in Equation (3.77).

Next, consider a beam which is loaded and supported as shown in Figure 4.18. Suppose we are interested in finding the beam rotation at each of the pin supports and the deflection and rotation at the midpoint of the beam. We can proceed as follows: Let the beam be divided into 3 elements as shown in Figure 4.19. If L is the overall length of the beam, let ℓ be $L/4$. (Note that the element nodes are strategically placed at the supports and at the midpoint of the beam.) From Equation (4.162) the element stiffness matrices are found to be

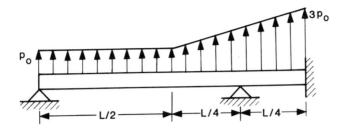

Figure 4.18 Multiply Supported Beam with Distributed Loading

$$
\overset{(1)}{[k]} = EI \begin{bmatrix}
3/2\ell^3 & 3/2\ell^2 & -3/2\ell^3 & 3/2\ell^2 \\
3/2\ell^2 & 2/\ell & -3/2\ell^2 & 1/\ell \\
-3/2\ell^3 & 3/2\ell^2 & 3/2\ell^3 & -3/2\ell^2 \\
3/2\ell^2 & 1/\ell & -3/2\ell^2 & 2/\ell
\end{bmatrix} \quad (4.172)
$$

$$
\overset{(2)}{[k]} = \overset{(3)}{[k]} = EI \begin{bmatrix}
12/\ell^3 & 6/\ell^2 & -12/\ell^3 & 6/\ell^2 \\
6/\ell^2 & 4/\ell & -6/\ell^2 & 2/\ell \\
-12/\ell^3 & -6/\ell^2 & 12/\ell^3 & -6/\ell^2 \\
6/\ell^2 & 2/\ell & -6/\ell^2 & 4/\ell
\end{bmatrix} \quad (4.173)
$$

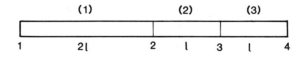

Figure 4.19 Finite Element Model of the Beam of Figure 4.18

From Equations (4.170) and (4.171), the element "fixed-end" load arrays are found to be

$$\{\overset{(1)}{p}\} = \begin{bmatrix} p_0 \ell \\ (1/3)p_0 \ell^2 \\ p_0 \ell \\ -(1/3)p_0 \ell^2 \end{bmatrix} \qquad (4.174)$$

$$\{\overset{(2)}{p}\} = \begin{bmatrix} (13/20)p_0 \ell \\ (7/60)p_0 \ell^2 \\ (17/20)p_0 \ell \\ -(2/15)p_0 \ell^2 \end{bmatrix} \qquad (4.175)$$

$$\{\overset{(3)}{p}\} = \begin{bmatrix} (23/20)p_0 \ell \\ (1/5)p_0 \ell^2 \\ (27/20)p_0 \ell \\ -(13/60)p_0 \ell^2 \end{bmatrix} \qquad (4.176)$$

The global stiffness matrix $[K]$ is an 8×8 array which may be obtained in the usual manner by superposing the element stiffness arrays. Specifically, $[K]$ may be visualized as

$$[K] = \begin{bmatrix} \begin{bmatrix} \overset{(1)}{k} \end{bmatrix} & & 0 \\ & \begin{bmatrix} \overset{(2)}{k} \end{bmatrix} & \\ 0 & & \begin{bmatrix} \overset{(3)}{k} \end{bmatrix} \end{bmatrix} \qquad (4.177)$$

By carrying out the indicated summations $[K]$ is found to be

$$[K] = EI \begin{bmatrix}
(3/2\ell^3) & (3/2\ell^2) & (-3/2\ell^3) & (3/2\ell^2) & 0 & 0 & 0 & 0 \\
(3/2\ell^2) & (2/\ell) & (-3/2\ell^2) & (1/\ell) & 0 & 0 & 0 & 0 \\
(-3/2\ell^3) & (-3/2\ell^2) & (27/2\ell^3) & (9/2\ell^2) & (-12/\ell^3) & (6/\ell^2) & 0 & 0 \\
(3/2\ell^2) & (1/\ell) & (9/2\ell^2) & (6/\ell) & (-6/\ell^2) & (2/\ell) & 0 & 0 \\
0 & 0 & (-12/\ell^3) & (-6/\ell^2) & (24/\ell^3) & 0 & (-12/\ell^3) & (6/\ell^2) \\
0 & 0 & (6/\ell^2) & (2/\ell) & 0 & (8/\ell) & (-6/\ell^2) & (2/\ell) \\
0 & 0 & 0 & 0 & (-12/\ell^3) & (-6/\ell^2) & (12/\ell^3) & (-6/\ell^2) \\
0 & 0 & 0 & 0 & (6/\ell^2) & (2/\ell) & (-6/\ell^2) & (4/\ell)
\end{bmatrix} \quad (4.178)$$

The governing equations are of the usual form, that is,

$$[K] \{U\} = \{L\} + \{R\} \tag{4.179}$$

where from Equations (4.174), (4.175), and (4.176) the transpose of the "fixed-end" load array is

$$[L]^T = [p_0 \ell, (1/3)p_0 \ell^2, (33/20)p_0 \ell, (-13/60)p_0 \ell^2,$$

$$2p_0 \ell, (1/15)p_0 \ell^2, (27/20)p_0 \ell, (-13/60)p_0 \ell^2] \tag{4.180}$$

From Figures 4.18 and 4.19 the transposes of the displacement and reaction arrays are found to be

$$[U]^T = [0, \theta_1, V_2, \theta_2, 0, \theta_3, 0, 0] \tag{4.181}$$

and

$$[R]^T = [R_{1Y}, 0, 0, 0, R_{3Y}, 0, R_{4Y}, M_4] \tag{4.182}$$

The matrix governing equations are thus obtained by substituting Equations (4.178), (4.180), (4.181), and (4.182) into Equation (4.179). We may obtain the unknown deflection and rotation of Equation (4.181) by examining the second, third, fourth, and sixth of these equations. These are

$$(2/\ell)\theta_1 - (3/2\ell^2)V_2 + (1/\ell)\theta_2 = (1/3 p_0 \ell^2/EI$$

$$-(3/2\ell^2)\theta_1 + (27/2\ell^3)V_2 + (9/2\ell^2)\theta_2 + (6/\ell^2)\theta_3$$

$$= (33/20)p_0 \ell/EI$$

$$(1/\ell)\theta_1 + (9/2\ell^2)V_2 + (6/\ell)\theta_2 + (2/\ell)\theta_3 = (-13/60)p_0 \ell^2/EI$$

$$(6/\ell^2)V_2 + (2/\ell)\theta_2 + (8/\ell)\theta_3 = (1/15)p_0 \ell^2/EI$$

$$\tag{4.183}$$

Solving for θ_1, V_2, θ_2, and θ_3 leads to the results

$$\theta_1 = (154/225)p_0 \ell^3/EI = (77/7200)p_0 L^3/EI$$

$$V_2 = (172/405)p_0 \ell^4/EI = (43/25920)p_0 L^3/EI$$

$$\theta_2 = -(269/675)p_0 \ell^3/EI = -(269/43200)p_0 L^3/EI$$

$$\theta_3 = -(379/1800)p_0 \ell^3/EI = -(379/115200)p_0 L^3/EI$$

$$\tag{4.184}$$

4.14 CONCLUDING REMARKS

The procedure for obtaining the element stiffness matrix in this chapter is based upon a variational formulation leading to a minimization of a functional. This procedure uses the governing differential equation of the system as opposed to the strictly physical approach we used in Chapters 2 and 3. As a consequence, the method of this chapter is more versatile as seen in the applications with beam problems. That is, by using interpolation functions such as the Hermitian polynomials which are "natural" solutions of the governing differential equations, the computations are greatly simplified. Therefore, the range and complexity of the problems which may be addressed are increased.

These ideas are further illustrated and developed in the following chapter on two-dimensional elasticity and heat transfer.

PROBLEMS

Section 4.2

4.1 Verify the approximation of Equation (4.8). That is, show that the approximation is exact at X_e. Next, show that between X_e and X_{e+1} the approximation is linear.

4.2 Use Equation (4.8) to find a piecewise linear approximation to the parabola: $y = 2x - x^2/2$ using 4 elements. Sketch the function and its approximation.

Section 4.3

4.3 Using Equation (4.16), verify Equation (4.17) for the 4 element division represented in Figure 4.8.

Section 4.4

    4.4 Show, by referring to Equation (4.12) that $\sum\limits_{j=1}^{2} \overset{(e)}{y_j} \overset{(e)}{N_j}(X)$
may be expanded into the form $\overset{(e)}{C_1} + \overset{(e)}{C_2}x$ where $\overset{(e)}{C_1}$ and $\overset{(e)}{C_2}$ are

$$\overset{(e)}{C_1} = [\overset{(e)}{y_1}\overset{(e)}{x_2} - \overset{(e)}{y_2}\overset{(e)}{x_1}]/\ell$$

and

$$\overset{(e)}{C_2} = [\overset{(e)}{y_2} - \overset{(e)}{y_1}]/\ell$$

Section 4.5

    4.5 Using Equations (4.53) verify Equation (4.54).

Section 4.6

    4.6 Verify Equations (4.62) and (4.63) by using Equations (4.60) and (4.61). Check the result by comparing with Equation (4.59).

    4.7 Repeat Problem 4.6 for the linear interpolation functions and the incidence matrices of Equation (4.16). Check the results with Figure 4.6.

    4.8 Develop an analog of the pyramid functions for the quadratic interpolation functions of Section 4.5. That is, let

$$\overset{(k-1)}{\phi}(x) = \sum_{e=1}^{n} \sum_{j=1}^{3} \overset{(e)}{\lambda_{kj}} \overset{(e)}{N_{kj}}(x) \tag{4.64}$$

[Compare with Equation (4.17)] Show that for a 3-element partitioning of an interval as in Figures 4.9 to 4.12, that the $\{\Phi(x)\}$ array becomes

$$\{\Phi(x)\} = \left\{ \begin{array}{c} \overset{(1)}{N}_1 \\[1em] \overset{(1)}{N}_2 \\[1em] \overset{(1)}{N}_3 + \overset{(2)}{N}_1 \\[1em] \overset{(2)}{N}_2 \\[1em] \overset{(2)}{N}_3 + \overset{(3)}{N}_1 \\[1em] \overset{(3)}{N}_2 \\[1em] \overset{(3)}{N}_3 \end{array} \right\} \tag{4.65}$$

4.9   Sketch the $\overset{(k-1)}{\phi}(x)$ functions developed in Problem 4.8.

4.10   Solve Equation (4.63) for $\{Y\}$ by multiplying both sides by $[\overset{(e)}{\Lambda}]$. Discuss the results.

Section 4.7

4.11   Verify Equation (4.79).

4.12   Verify Equation (4.82).

4.13   Verify Equation (4.83).

Section 4.8

4.14 Show that the solution of the boundary value problem

$$\frac{d^2y}{dx^2} - y = f_0$$

where

$$y(0) = 1 - f_0 \quad \text{and} \quad y(4) = e^4 - f_0$$

is

$$y = e^x - f_0$$

4.15 Find an approximate solution to the boundary value problem of Problem 4.14 by using the finite element method as illustrated in the foregoing section, but by using 8 elements or subintervals instead of 4. Compare the results with the exact solution and with Equation (4.107).

4.16 Use the finite element method to find an approximate solution to the boundary value problem of Equations (4.1) and (4.2) of Section 4.1. That is, let the right-hand side and the boundary conditions have arbitrary values as opposed to the specific values used in Section 4.8 and Problem 4.14. Check the result against some specific cases.

Section 4.11

4.17 Selectively verify some of the integrals of Equation (4.130).

4.18 Verify Equation (4.140).

4.19 Verify Equation (4.141).

Section 4.12

4.20 Using the procedure that was used to develop Equation (4.166), verify the results of Equation (4.167).

Section 4.13

4.21 Verify Equations (4.70) and (4.71).

4.22 Verify Equations (4.72) and (4.73).

4.23 Verify Equation (4.178).

4.24 Verify Equation (4.180).

4.25 Using the results of Equation (4.184), use the first, fifth, seventh, and eighth of the governing equations (4.79) to obtain the unknown reactions of Equation (4.182).

Answer: $R_{1Y} = -(281/5400)p_0 L,$ $R_{3Y} = -(913/1350)p_0 L$

$R_{4Y} = (379/1200)p_0 L,$ $M_4 = -(379/14400)p_0 L^2$

5

Introduction to Two-Dimensional Analysis

5.1 INTRODUCTION

In this chapter we will apply the finite element procedures in the analysis of two-dimensional boundary value problems. As in the earlier chapters, our analysis will be restricted to elementary problems so that the basic principles and procedures of the method will become evident. Since many of these procedures are the same as those discussed earlier (for example, the variational techniques, the development of stiffness matrices, and the assembly procedure), we will focus primarily upon those concepts which are new or are characteristics of two-dimensional problems.

As before, we will develop our analysis through the solution of relatively simple example problems. We will begin with an examination of steady state heat conduction in a plane. Later we will examine two-dimensional plane stress/plane strain problem formulations.

5.2 THE STEADY-STATE HEAT CONDUCTION PROBLEM

The two-dimensional steady-state heat conduction problem might be stated as follows: Consider the region R bounded by the curve C in the X-Y plane as shown in Figure 5.1. Then in R, and on C, find the temperature function $T(x, y)$ which satisfies the partial differential equation

$$\nabla \cdot (k\nabla T) = f \text{ in } R \tag{5.1}$$

subject to the boundary conditions

$$T = g \text{ on } C_1 \tag{5.2}$$

and

$$\nabla T \cdot n = h \text{ on } C_2 \tag{5.3}$$

where k, f, g, and h are given functions of x and y and where the boundary curve C is divided into the parts C_1 and C_2. [Regarding notation, ∇ is the two-dimensional vector differential operator

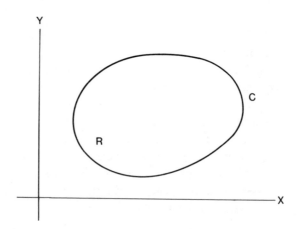

Figure 5.1   Heat Conduction Region

$$\nabla(\;\;) = i\partial(\;\;)/\partial x + j\partial(\;\;)/\partial y \tag{5.4}$$

and n is a unit vector normal to C and directed out of R.)

In Equation (5.1), f $[= f(x, y)]$ represents "heat sources" or "heat sinks" within R. f is thus analogous to a "forcing function." Similarly, in Equation (5.3), h represents a forced heat conduction on the boundary of C. In many cases of practical interest, f and h are zero. Also, the conduction coefficient k is usually assumed to be constant. In these cases Equations (5.1), (5.2), and (5.3) take the simple form

$$\nabla \cdot \nabla T = \nabla^2 T = 0 \text{ in } R \tag{5.5}$$

and

$$T = g \text{ on } C_1 \quad \text{and} \quad \nabla T \cdot n = 0 \text{ on } C_2 \tag{5.6}$$

To convert Equation (5.1), together with the boundary conditions of Equations (5.2) and (5.3), into a variational formulation, we can proceed as before by multiplying Equation (5.1) by a variation δT in T and by integrating over the region R. Recall that in our earlier variational formulations (for example, with beam theory) we obtained a functional or integral to be minimized by integrating by parts. The analog of this procedure in two and three dimensions is to use an integral theorem such as the divergence theorem. For the region R this may be written

$$\int_R \nabla \cdot v dR = \int_C v \cdot n dC \tag{5.7}$$

where v is any vector. To effectively use this theorem, it is also helpful to recall the vector identity

$$\nabla \cdot (sv) = (\nabla s) \cdot v + s(\nabla \cdot v) \tag{5.8}$$

where s is a scalar. If s is δT and v is ∇T, then Equation (5.8) may be written

$$\nabla \cdot (\delta T \nabla T) = (\nabla \delta T) \cdot \nabla T + \delta T \nabla^2 T = (\tfrac{1}{2})\delta(\nabla T)^2 + \delta T \nabla^2 T \tag{5.9}$$

By multiplying the left side of Equation (5.1) by δT, by assuming k to be constant, by integrating over R, and by using Equation (5.9), we obtain

$$k \int_R \delta T \nabla^2 T dR = k \int_R \nabla \cdot (\delta T \nabla T) dR - \frac{k}{2} \int_R \delta (\nabla T)^2 dR \tag{5.10}$$

By using the divergence theorem with the first integral on the right side, this becomes

$$k \int_R \delta T \nabla^2 T dR = - \frac{k}{2} \int_R \delta (\nabla T)^2 dR + k \int_C \delta T \nabla T \cdot n dC \tag{5.11}$$

If we require that δT be zero on C_1 and by assuming that h is zero, the last integral in Equation (5.11) vanishes. Finally, multiplying the right side of Equation (5.1) by δT and integrating over R leads to the simple result

$$\int_R \delta T f dR = \delta \int_R f T dR \tag{5.12}$$

Hence, by combining Equations (5.11) and (5.12), we see that the variational formulation of the boundary value problem of Equation (5.1) with k constant is

$$\delta I = 0 \tag{5.13}$$

where I is

$$I = \int_R [k(\nabla T)^2 + 2fT] dR \tag{5.14}$$

where $T = g$ on C_1 [Equation (5.2)] and $h = 0$ [Equation (5.3)].

5.3 FINITE ELEMENT APPROXIMATION: TRIANGULAR ELEMENTS

Consider again the heat conduction region of Figure 5.1. Let the region be divided into triangles as shown in Figure 5.2. Let these triangles be "finite elements" of the region. Our approach in seeking

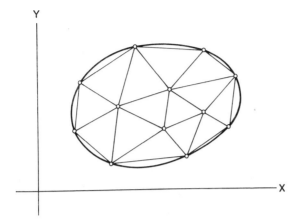

Figure 5.2   Triangular Element Representation
of the Heat Conduction Region

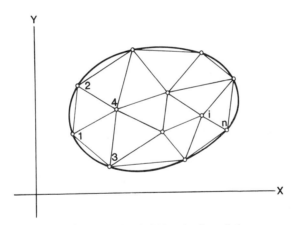

Figure 5.3   Global Numbering of the
Element Nodes of Figure 5.2

a solution to the heat conduction problem of Equations (5.1) to
(5.3) or (5.13) will be to represent the temperature function $T(x, y)$
throughout R by its values at the vertices or "nodes" of the triangles.
To this end, let the vertices of the triangles be numbered globally
throughout R as shown in Figure 5.3. Then, the temperature distribution in R will be represented by the array

$$\{T\}^T = \{T_1, T_2, T_3, \ldots, T_n\} \tag{5.15}$$

At points of R which are not nodes, that is, in the interior and on the edges of the triangular elements, we will assume that the temperature varies linearly over the elements and between the nodes.*

One way of visualizing this representation is to introduce "pyramid functions," similar to those used in Chapter 4, with the property of having the value 1 at a given node and the value 0 at all other nodes. Also, let this function vary linearly from 1 to 0 over all triangles or elements with vertices at the given node. Specifically, let $\phi_i(x, y)$ be the pyramid function associated with node i. Then, ϕ_i has the property

$$\phi_i(x_i, y_i) = 1 \quad \text{and} \quad \phi_i(x_j, y_j) = 0 \tag{5.16}$$

where (x_j, y_j) are the coordinates of node j. Then, the temperature distribution throughout the region R may be represented by the expression

$$T(x, y) = \sum_{i=1}^{n} T_i \phi_i(x, y) \tag{5.17}$$

5.4 LOCAL OR ELEMENT TEMPERATURE REPRESENTATION

To develop the details of the above temperature representation at the element level, consider a typical triangular element (e) as shown in Figure 5.4. Let the vertices or local nodes of the triangle be labeled "1," "2," and "3," as shown. Let the X, Y coordinates of these nodes be: (x_i, y_i), $i = 1, 2, 3$, as shown in Figure 5.5.

To develop the desired linear temperature distribution over the element, let us introduce three linear element interpolation functions, or "shape functions" $N_i^{(e)}(x, y)$, $i = 1, 2, 3$, defined with the following properties:

---

*This linearity assumption can, of course, be modified to include higher order (for example, quadratic) approximation.

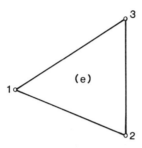

Figure 5.4 A Typical Triangular Element

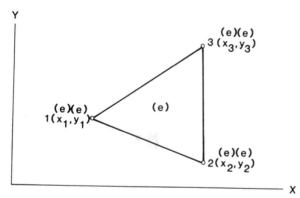

Figure 5.5 A Typical Triangular Element
with Nodal Coordinates

1. Let $\overset{(e)}{N_i}(x, y)$ be linear in both x and y on (e).

2. Let $\overset{(e)}{N_i}(x, y)$ be zero at all points in R except those of (e).

3. Let $\overset{(e)}{N_i}(x, y)$ have the value 1 at node i and the values 0 at nodes j, $j \neq i$.

Hence, from Equation (5.17), the temperature representation on (e) may be expressed as

$$\overset{(e)}{T}(x, y) = \overset{(e)}{T_1} \overset{(e)}{N_1}(x, y) + \overset{(e)}{T_2} \overset{(e)}{N_2}(x, y) + \overset{(e)}{T_3} \overset{(e)}{N_3}(x, y) \tag{5.18}$$

To determine explicit expressions for $\overset{(e)}{N_i}(x, y)$, let $\overset{(e)}{N_i}(x, y)$ be written in the form

$$\overset{(e)}{N_i}(x, y) = \overset{(e)}{a_i} + \overset{(e)}{b_i}x + \overset{(e)}{c_i}y \quad (i = 1, 2, 3) \tag{5.19}$$

where $\overset{(e)}{a_i}$, $\overset{(e)}{b_i}$, and $\overset{(e)}{c_i}$ are constant coefficients to be determined from the nodal coordinates. To determine these coefficients, consider first $\overset{(e)}{N_1}$: From property 3 above we can write the equations*

$$1 = a_1 + b_1 x_1 + c_1 y_1$$

$$0 = a_1 + b_1 x_2 + c_1 y_2 \tag{5.20}$$

$$0 = a_1 + b_1 x_3 + c_1 y_3$$

These equations form a consistent set of linear equations for the coefficients a_1, b_1, and c_1. To efficiently solve this system, let the matrix of coefficients be $[A]$; that is, let

$$[A] = \begin{bmatrix} 1 & x_1 & y_1 \\ 1 & x_2 & y_2 \\ 1 & x_3 & y_3 \end{bmatrix} \tag{5.21}$$

Then the system of Equations (5.20) may be written in the matrix form

$$\begin{bmatrix} 1 \\ 0 \\ 0 \end{bmatrix} = [A] \begin{bmatrix} a_1 \\ b_1 \\ c_1 \end{bmatrix} \tag{5.22}$$

---

*Regarding notation, we will delete the superscript (e) for simplicity where it is clear that we are working with element parameters.

Hence, the solution may be expressed as

$$
\begin{bmatrix} a_1 \\ b_1 \\ c_1 \end{bmatrix} = [A]^{-1} \begin{bmatrix} 1 \\ 0 \\ 0 \end{bmatrix}
\tag{5.23}
$$

Similarly, for the interpolation function N_2 and N_3 we obtain

$$
\begin{bmatrix} a_2 \\ b_2 \\ c_2 \end{bmatrix} = [A]^{-1} \begin{bmatrix} 0 \\ 1 \\ 0 \end{bmatrix} \quad \text{and} \quad \begin{bmatrix} a_3 \\ b_3 \\ c_3 \end{bmatrix} = [A]^{-1} \begin{bmatrix} 0 \\ 0 \\ 1 \end{bmatrix}
\tag{5.24}
$$

From Equation (5.19), $N_i(x, y)$ may be written in the matrix form

$$
N_i(x, y) = [a_i \ b_i \ c_i] \begin{bmatrix} 1 \\ x \\ y \end{bmatrix}
\tag{5.25}
$$

or as

$$
\begin{bmatrix} N_1 \\ N_2 \\ N_3 \end{bmatrix} = \begin{bmatrix} a_1 & b_1 & c_1 \\ a_2 & b_2 & c_2 \\ a_3 & b_3 & c_3 \end{bmatrix} \begin{bmatrix} 1 \\ x \\ y \end{bmatrix}
\tag{5.26}
$$

By taking the transpose, this becomes

$$
[N_1 \ N_2 \ N_3] = [1 \ x \ y] \begin{bmatrix} a_1 & a_2 & a_3 \\ b_1 & b_2 & b_3 \\ c_1 & c_2 & c_3 \end{bmatrix}
\tag{5.27}
$$

From Equations (5.23) and (5.24) this in turn may be written

$$[N_1 \ N_2 \ N_3] = [1 \ x \ y] \ [A]^{-1} \begin{bmatrix} 1 & 0 & 0 \\ 0 & 1 & 0 \\ 0 & 0 & 1 \end{bmatrix}$$

$$= [1 \ x \ y] \ [A]^{-1} \ [I] = [1 \ x \ y] \ [A]^{-1} \quad (5.28)$$

Consider next the computation of $[A]^{-1}$: Recall from matrix theory (see Chapter 1, Section 1.2) that the inverse of a nonsingular square matrix may be expressed as the transpose of the cofactor matrix divided by the determinant. Hence, from Equation (5.21), $[A]^{-1}$ is found to be

$$[A]^{-1} = (1/\det[A]) \begin{bmatrix} (x_2 y_3 - x_3 y_2) & (y_2 - y_3) & (x_3 - x_2) \\ (x_3 y_1 - x_1 y_3) & (y_3 - y_1) & (x_1 - x_3) \\ (x_1 y_2 - x_2 y_1) & (y_1 - y_2) & (x_2 - x_1) \end{bmatrix}^T$$

$$(5.29)$$

Let a be the area of the triangular element (e) and let Δ be $\det[A]$. (Note the slight change in notation from that suggested in Chapter 1.) Then, interestingly, it can be shown (see Problem 5.2) that

$$\det[A] = \Delta = 2a \quad (5.30)$$

Hence, $[A]^{-1}$ may be written

$$[A]^{-1} = (1/2a) \begin{bmatrix} (x_2 y_3 - x_3 y_2) & (x_3 y_1 - x_1 y_3) & (x_1 y_2 - x_2 y_1) \\ (y_2 - y_3) & (y_3 - y_1) & (y_1 - y_2) \\ (x_3 - x_2) & (x_1 - x_3) & (x_2 - x_1) \end{bmatrix}$$

$$(5.31)$$

Then finally, by substituting from Equation (5.31) into Equation (5.38), the element interpolation shape functions are found to be

$$N_1(x, y) = (\tfrac{1}{2}a)\, [(x_2 y_3 - x_3 y_2) + x(y_2 - y_3) + y(x_3 - x_2)]$$
$$(5.32)$$

$$N_2(x, y) = (\tfrac{1}{2}a)\, [(x_3 y_1 - x_1 y_3) + x(y_3 - y_1) + y(x_1 - x_3)]$$
$$(5.33)$$

$$N_3(x, y) = (\tfrac{1}{2}a)\, [(x_1 y_2 - x_2 y_1) + x(y_1 - y_2) + y(x_2 - x_1)]$$
$$(5.34)$$

where $N_i(x, y) = \overset{(e)}{N_i}(x, y)$ are zero except on (e).

5.5 ELEMENT TEMPERATURE GRADIENTS

From Equation (5.18) the temperature distribution over a typical element (e) is given by the expression

$$\overset{(e)}{T}(x, y) = \overset{(e)}{T_1}\, \overset{(e)}{N_1}(x, y) + \overset{(e)}{T_2}\, \overset{(e)}{N_2}(x, y) + \overset{(e)}{T_3}\, \overset{(e)}{N_3}(x, y) \qquad (5.35)$$

In matrix form this may be written

$$\overset{(e)}{T}(x, y) = [\overset{(e)}{N_1}\ \overset{(e)}{N_2}\ \overset{(e)}{N_3}]
\begin{bmatrix} \overset{(e)}{T_1} \\ \overset{(e)}{T_2} \\ \overset{(e)}{T_3} \end{bmatrix}
= \{\overset{(e)}{N}(x, y)\}^T\, \{\overset{(e)}{T}\} = \{\overset{(e)}{T}\}^T\, \{\overset{(e)}{N}(x, y)\}$$
$$(5.36)$$

From this expression, it is seen that in addition to providing the desired linear representation over element (e), the interpolation or shape functions $\overset{(e)}{N_i}(x, y)$ also provide a means of separating the dependence of the temperature upon x and y and the dependence upon its nodal values.

To determine the element nodal temperature values $\overset{(e)}{T_1}$, $\overset{(e)}{T_2}$, and $\overset{(e)}{T_3}$ which will minimize the integral of Equation (5.14) (and thus ultimately to determine the global temperature distribution), it is necessary to calculate the element temperature gradient: $\overset{(e)}{\nabla T}$. Recall that in matrix form the gradient may be expressed as

$$\overset{(e)}{\nabla T} = \begin{bmatrix} \partial \overset{(e)}{T}/\partial x \\ \partial \overset{(e)}{T}/\partial y \end{bmatrix} \tag{5.37}$$

Hence, from Equations (5.36) and (5.27), $\overset{(e)}{\nabla T}$, becomes

$$\overset{(e)}{\nabla T} = \begin{bmatrix} \partial \overset{(e)}{N_1}/\partial x & \partial \overset{(e)}{N_2}/\partial x & \partial \overset{(e)}{N_3}/\partial x \\ \partial \overset{(e)}{N_1}/\partial y & \partial \overset{(e)}{N_2}/\partial y & \partial \overset{(e)}{N_3}/\partial y \end{bmatrix} \begin{bmatrix} \overset{(e)}{T_1} \\ \overset{(e)}{T_2} \\ \overset{(e)}{T_3} \end{bmatrix}$$

$$= \begin{bmatrix} b_1 & b_2 & b_3 \\ c_1 & c_2 & c_3 \end{bmatrix} \begin{bmatrix} \overset{(e)}{T_1} \\ \overset{(e)}{T_2} \\ \overset{(e)}{T_3} \end{bmatrix} \tag{5.38}$$

Let the coefficient matrix be called $\overset{(e)}{[B]}$. That is, let

$$\overset{(e)}{[B]} = \begin{bmatrix} b_1 & b_2 & b_3 \\ c_1 & c_2 & c_3 \end{bmatrix} \tag{5.39}$$

Then, $\overset{(e)}{\nabla T}$ might be written in the compact form

$$\overset{(e)}{\nabla T} = \overset{(e)}{[B]}\overset{(e)}{\{T\}} \tag{5.40}$$

Hence, $(\overset{(e)}{\nabla T})^2$ becomes

$$(\overset{(e)}{\nabla T})^2 = \overset{(e)}{\{T\}}^T \overset{(e)}{[B]}^T \overset{(e)}{[B]}\overset{(e)}{\{T\}} \tag{5.41}$$

Next, consider the "forcing function" term fT in the integral of Equation (5.14): Let fT be written on (e) as: $\overset{(e)}{f}\ \overset{(e)}{T}$. $\overset{(e)}{T}$ is given by Equation (5.36). Let $\overset{(e)}{f}$ be expressed in the similar form

$$\overset{(e)}{f} = \{\overset{(e)}{N}(x,\,y)\}^T\{\overset{(e)}{f}\} = \{\overset{(e)}{f}\}^T\{\overset{(e)}{N}(x,\,y)\} \tag{5.42}$$

where $\{\overset{(e)}{f}\}^T$ is the array $[\overset{(e)}{f_1}\ \overset{(e)}{f_2}\ \overset{(e)}{f_3}]$ where $\overset{(e)}{f_i}$ $(i = 1,\ 2,\ 3)$ are the values of $f(x,\ y)$ at the element nodes. Hence, on (e) the product $\overset{(e)}{f}\ \overset{(e)}{T}$ may be expressed as

$$\overset{(e)}{f}\ \overset{(e)}{T} = \{\overset{(e)}{f}\}^T\{\overset{(e)}{N}\}\,\{\overset{(e)}{N}\}^T\{\overset{(e)}{T}\} \tag{5.43}$$

5.6 ELEMENT STIFFNESS MATRIX

Consider again the variational formulation of our heat conduction problem. From Equations (5.13) and (5.14) we have

$$\delta I = 0 \quad \text{where} \quad I = \int_R [k(\nabla T)^2 + 2fT]\ dR \tag{5.44}$$

If we divide the region R into N elements as shown in Figure 5.6, I may be written in the form

$$I = \sum_{e=1}^{N} \overset{(e)}{I} \tag{5.45}$$

where $\overset{(e)}{I}$ is

$$\overset{(e)}{I} = (\tfrac{1}{2}) \int_{(e)} [k(\nabla \overset{(e)}{T})^2 + 2\overset{(e)}{f}\ \overset{(e)}{T}]\ dR \tag{5.46}$$

Consider the first term in this integral: By substituting from Equation (5.41) this term may be written

$$\int_{(e)} k(\nabla \overset{(e)}{T})^2 dR = k \int_{(e)} (\{\overset{(e)}{T}\}^T[\overset{(e)}{B}]^T[\overset{(e)}{B}]\{\overset{(e)}{T}\})\ dR \tag{5.47}$$

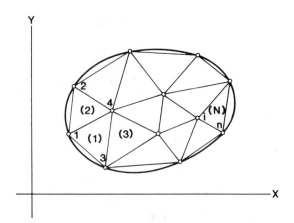

Figure 5.6 Finite Element Division of
the Heat Conduction Region

However, from Equation (5.39) we see that $[B]$ is constant on (e). Hence, by recalling that k is also constant, the integral may be written

$$\int_{(e)} k(\overset{(e)}{\nabla T})^2 dR = k \{\overset{(e)}{T}\}^T [\overset{(e)}{B}]^T [\overset{(e)}{B}] \{\overset{(e)}{T}\} \overset{(e)}{a} \tag{5.48}$$

where $\overset{(e)}{a}$ is the area of the triangular element (e).

Consider the second term in Equation (5.46): Using Equation (5.43) this may be written

$$\int_{(e)} 2\overset{(e)}{f} \overset{(e)}{T} dR = 2\int_{(e)} \{\overset{(e)}{f}\}^T \{\overset{(e)}{N}\} \{\overset{(e)}{N}\}^T \{\overset{(e)}{T}\} dR = 2\{\overset{(e)}{g}\}^T \{\overset{(e)}{T}\} \tag{5.49}$$

where $\{\overset{(e)}{g}\}^T$ is defined

$$\{\overset{(e)}{g}\}^T = \{\overset{(e)}{f}\}^T \int_{(e)} \{\overset{(e)}{N}\} \{\overset{(e)}{N}\}^T dR \tag{5.50}$$

Hence, from Equations (5.48) and (5.49), $\overset{(e)}{I}$ may be written in the form

$$\overset{(e)}{I} = (\tfrac{1}{2}) \{\overset{(e)}{T}\}^T [\overset{(e)}{B}]^T [\overset{(e)}{B}] \{\overset{(e)}{T}\} \overset{(e)}{a} + \{\overset{(e)}{g}\}^T \{\overset{(e)}{T}\}$$

$$= (\tfrac{1}{2}) \{\overset{(e)}{T}\}^T [\overset{(e)}{k}] \{\overset{(e)}{T}\} + \{\overset{(e)}{g}\}^T \{\overset{(e)}{T}\} \qquad (5.51)$$

where $[\overset{(e)}{k}]$ is the "element stiffness matrix" which is defined

$$[\overset{(e)}{k}] = k [\overset{(e)}{B}]^T [\overset{(e)}{B}] \overset{(e)}{a} \qquad (5.52)$$

From Equation (5.39) $[\overset{(e)}{k}]$ may be written in terms of b_i and c_i ($i = 1, 2, 3$)

$$[\overset{(e)}{k}] = \overset{(e)}{k}a \begin{bmatrix} (b_1^2 + c_1^2) & (b_1 b_2 + c_1 c_2) & (b_1 b_3 + c_1 c_3) \\ (b_2 b_1 + c_2 c_1) & (b_2^2 + c_2^2) & (b_2 b_3 + c_2 c_3) \\ (b_3 b_1 + c_3 c_1) & (b_3 b_2 + c_3 c_2) & (b_3^2 + c_3^2) \end{bmatrix}$$

$$(5.53)$$

5.7 ELEMENT FORCING FUNCTION, AREA COORDINATES

Before discussing the assembly procedure, the development of the global stiffness matrix, and the development of the global forcing function, it is helpful first to consider the integral in Equation (5.50): From Equations (5.26) and (5.27) the integrand $\{\overset{(e)}{N}\}\{\overset{(e)}{N}\}^T$ might be written

$$\{\overset{(e)}{N}\}\{\overset{(e)}{N}\}^T = [A]^{-1} \begin{bmatrix} 1 \\ x \\ y \end{bmatrix} [1 \; x \; y] \, [A^T]^{-1}$$

$$= [A]^{-1} \begin{bmatrix} 1 & x & y \\ x & x^2 & xy \\ y & xy & y^2 \end{bmatrix} [A^T]^{-1} \qquad (5.54)$$

where $[A]^{-1}$ is the array

$$[A]^{-1} = \begin{bmatrix} a_1 & b_1 & c_1 \\ a_2 & b_2 & c_2 \\ a_3 & b_3 & c_3 \end{bmatrix} \tag{5.55}$$

Since $[A]^{-1}$ is independent of x and y, Equation (5.50) may thus be written

$$\{\overset{(e)}{g}\}^T = 2 \{f\}^T [A]^{-1} \left(\int_{(e)} \begin{bmatrix} 1 & x & y \\ x & x^2 & xy \\ y & xy & y^2 \end{bmatrix} dR \right) [A^T]^{-1} \tag{5.56}$$

or, by taking the tranpose as

$$\{\overset{(e)}{g}\} = 2 [A^T]^{-1} \left(\int_{(e)} \begin{bmatrix} 1 & x & y \\ x & x^2 & xy \\ y & xy & y^2 \end{bmatrix} dR \right) [A]^{-1} \{\overset{(e)}{f}\} \tag{5.57}$$

Hence, $\{\overset{(e)}{g}\}$ is determined as soon as the integral is evaluated.

Evaluation of the integral of Equation (5.56) or (5.57) can be very tedious and laborious depending upon the element triangle shape and orientation (see Problem 5.5). To avoid some of this computation, it is helpful to introduce the concept of "area coordinates": Consider the triangular element of Figure 5.7. Let P be any point in the interior or boundary of the triangle. By connecting P to the vertices 1, 2, 3 of the triangle three new triangles: $P23$, $P31$, and $P12$ are formed. Let the areas of these triangles be a_1, a_2, and a_3, respectively, as shown in Figure 5.7. From this definition, note the following:

If P is at 1, then $a_1 = 1$ and $a_2 = a_3 = 0$.
If P is at 2, then $a_2 = 1$ and $a_3 = a_1 = 0$.
If P is at 3, then $a_3 = 1$ and $a_1 = a_2 = 0$.

[Compare this with the properties of $\overset{(e)}{N_i}(x, y)$ in Section 5.4.]

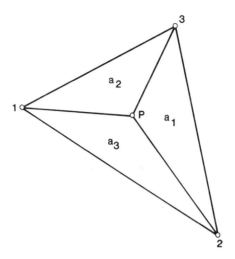

Figure 5.7 Subdivision of Triangular Element

Next, introduce three parameters α_1, α_2, and α_3 called the "area coordinates" of P defined

$$\alpha_1 = a_1/a \quad \alpha_2 = a_2/a \quad \text{and} \quad \alpha_3 = a_3/a \tag{5.58}$$

where a is the area of the original triangle: 123. Hence, we have

$$\alpha_1 + \alpha_2 + \alpha_3 = 1 \tag{5.59}$$

This mean, of course, that α_1, α_2, and α_3 are not independent. Indeed, only two of the three are independent.

As in the foregoing sections, let the coordinates of the triangle vertices 1, 2, and 3 be: (x_1, y_1), (x_2, y_2), and (x_3, y_3). Then recall from Problem 5.3 (Section 5.4) that

$$a = (\tfrac{1}{2}) \begin{vmatrix} 1 & x_1 & y_1 \\ 1 & x_2 & y_2 \\ 1 & x_3 & y_3 \end{vmatrix} \tag{5.60}$$

Then by generalizing this result, a_1, a_2, and a_3 are found to be

$$a_1 = (\tfrac{1}{2}) \begin{vmatrix} 1 & x & y \\ 1 & x_2 & y_2 \\ 1 & x_3 & y_3 \end{vmatrix}, \quad a_2 = (\tfrac{1}{2}) \begin{vmatrix} 1 & x_1 & y_1 \\ 1 & x & y \\ 1 & x_3 & y_3 \end{vmatrix},$$

and

$$a_3 = (\tfrac{1}{2}) \begin{vmatrix} 1 & x_1 & y_1 \\ 1 & x_2 & y_2 \\ 1 & x & y \end{vmatrix} \tag{5.61}$$

By expanding the determinants in this equation, the area coordinates $(\alpha_1, \alpha_2, \alpha_3)$ of P may be expressed in terms of the cartesian coordinates (x, y) of P as

$$\alpha_1 = (\tfrac{1}{2}a)\,[(x_2 y_3 - x_3 y_2) + x(y_2 - y_3) + y(x_3 - x_2)]$$

$$\alpha_2 = (\tfrac{1}{2}a)\,[(x_3 y_1 - x_1 y_3) + x(y_3 - y_1) + y(x_1 - x_3)] \tag{5.62}$$

$$\alpha_3 = (\tfrac{1}{2}a)\,[(x_1 y_2 - x_2 y_1) + x(y_1 - y_2) + y(x_2 - x_1)]$$

Recall from Equations (5.27) to (5.34) that the element interpolation functions $\overset{(e)}{N}$ may be written

$$\{\overset{(e)}{N}\} = [A]^{-1} \begin{bmatrix} 1 \\ x \\ y \end{bmatrix} \tag{5.63}$$

or more specifically

$$\begin{bmatrix} N_1 \\ N_2 \\ N_3 \end{bmatrix} = (\tfrac{1}{2}a) \begin{bmatrix} (x_2 y_3 - x_3 y_2) & (y_2 - y_3) & (x_3 - x_2) \\ (x_3 y_1 - x_1 y_3) & (y_3 - y_1) & (x_1 - x_2) \\ (x_1 y_2 - x_2 y_1) & (y_1 - y_2) & (x_2 - x_3) \end{bmatrix} \begin{bmatrix} 1 \\ x \\ y \end{bmatrix}$$

$$\tag{5.64}$$

Then by comparing Equations (5.62) and (5.64), we obtain the following interesting result:

$$\alpha_1 = N_1, \quad \alpha_2 = N_2, \quad \text{and} \quad \alpha_3 = N_3 \tag{5.65}$$

By using Equations (5.63), (5.64), and (5.65), we can reverse the above procedure and express the cartesian coordinates (x, y) of P in terms of the area coordinates $(\alpha_1, \alpha_2, \alpha_3)$ of P as follows: From Equations (5.63) and (5.65) we have

$$\begin{bmatrix} \alpha_1 \\ \alpha_2 \\ \alpha_3 \end{bmatrix} = [A^T]^{-1} \begin{bmatrix} 1 \\ x \\ y \end{bmatrix} \tag{5.66}$$

Then, by inverting, we have

$$\begin{bmatrix} 1 \\ x \\ y \end{bmatrix} = [A]^T \begin{bmatrix} \alpha_1 \\ \alpha_2 \\ \alpha_3 \end{bmatrix} = \begin{bmatrix} 1 & 1 & 1 \\ x_1 & x_2 & x_3 \\ y_1 & y_2 & y_3 \end{bmatrix} \begin{bmatrix} \alpha_1 \\ \alpha_2 \\ \alpha_3 \end{bmatrix} \tag{5.67}$$

Specifically, this means that

$$1 = \alpha_1 + \alpha_2 + \alpha_3 \tag{5.68}$$

$$x = \alpha_1 x_1 + \alpha_2 x_2 + \alpha_3 x_3 \tag{5.69}$$

and

$$y = \alpha_1 y_1 + \alpha_2 y_2 + \alpha_3 y_3 \tag{5.70}$$

The first of these expressions is a duplication of Equation (5.59). In view of Equation (5.65), the other two expressions may be written

$$x = x_1 N_1 + x_2 N_2 + x_3 N_3 \tag{5.71}$$

and

$$y = y_1 N_1 + y_2 N_2 + y_3 N_3 \tag{5.72}$$

Recall that our objective in introducing area coordinates is to be able to efficiently evaluate integrals as in Equation (5.57). It has been shown* that integrals of powers of the area coordinates may be evaluated by the simple expression

$$\int_{(e)} \alpha_1^m \, \alpha_2^n \, \alpha_3^p \, dR = \frac{m! \, n! \, p!}{(m+n+p+2)!} \, 2a \qquad (5.73)$$

where, as before, a is the triangle area and m, n, and p are integers.

Therefore, by returning to Equation (5.50), it is seen that $\{ \overset{(e)}{g} \}$ may be written in the form

$$\{ \overset{(e)}{g} \} = \left(\int_{(e)} \{ \overset{(e)}{N} \} \{ \overset{(e)}{N} \}^T \, dR \right) \{ \overset{(e)}{f} \} = \left(\int_{(e)} \begin{bmatrix} N_1^2 & N_1 N_2 & N_1 N_3 \\ N_2 N_1 & N_2^2 & N_2 N_3 \\ N_3 N_1 & N_3 N_2 & N_3^2 \end{bmatrix} dR \right) \{ \overset{(e)}{f} \}$$

$$(5.74)$$

In view of Equation (5.65) this may be written in terms of area coordinates as

$$\{ \overset{(e)}{g} \} = \left(\int_{(e)} \begin{bmatrix} \alpha_1^2 & \alpha_1 \alpha_2 & \alpha_1 \alpha_3 \\ \alpha_2 \alpha_1 & \alpha_2^2 & \alpha_2 \alpha_3 \\ \alpha_3 \alpha_1 & \alpha_3 \alpha_2 & \alpha_3^2 \end{bmatrix} dR \right) \{ \overset{(e)}{f} \} \qquad (5.75)$$

Hence, by using Equation (5.73) this becomes (see Problem 5.7)

$$\{ \overset{(e)}{g} \} = (a/12) \begin{bmatrix} 2 & 1 & 1 \\ 1 & 2 & 1 \\ 1 & 1 & 2 \end{bmatrix} \{ \overset{(e)}{f} \} \qquad (5.76)$$

---

*See, for example, M. A. Eisenberg and L. E. Malvern, On Finite Element Integration in Natural Coordinates, *International Journal for Numerical Methods in Engineering*, Vol. 7, 1973, pp. 574-575.

5.8 INCIDENCE MATRICES AND ASSEMBLY

To illustrate the assembly procedure, consider again the heat conduction region R and the typical element (e) as shown in Figure 5.8. The local nodes of the element (e): 1, 2, 3 correspond to global nodes: say, I, J, K of the region R. Therefore, the temperatures at the local nodes 1, 2, 3 are the same as the temperatures at the global nodes I, J, K, respectively. That is,

$$\overset{(e)}{T_1} = T_I, \quad \overset{(e)}{T_2} = T_J, \quad \overset{(e)}{T_3} = T_K \tag{5.77}$$

To formalize this relationship so that it can be applied generally, let us introduce incidence matrices as in the earlier chapters to relate the local and global temperature arrays. Specifically, let $\overset{(e)}{[\Lambda]}$ be an array of 1's and 0's such that

$$\overset{(e)}{\{T\}} = \overset{(e)}{[\Lambda]}^T \{T\} \tag{5.78}$$

where $\{T\}$ is the global temperature array. If the region R is divided into N elements with a total of n nodes, then there are N incidence matrices $\overset{(e)}{[\Lambda]}^T$ and they each have three rows and n columns. Moreover, all but three of the $3n$ entries of $\overset{(e)}{[\Lambda]}$ are *zero*. The nonzero

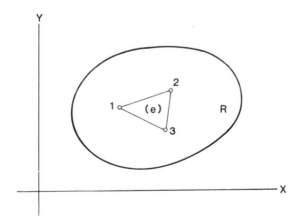

Figure 5.8 The Heat Conduction Region R
and a Typical Element (e)

entries (that is, the 1's) occur once in each row and in those columns I where the global node: I is the same as the local node: i. (This is illustrated in detail in the example problem in Section 5.10.)

The element forcing function array $\{\overset{(e)}{g}\}$ may be related to a global forcing function array $\{G\}$ in a similar manner. That is, consider the transpose of Equation (5.78):

$$\{\overset{(e)}{T}\}^T = \{T\}^T [\overset{(e)}{\Lambda}] \tag{5.79}$$

Then, let $\{\overset{(e)}{g}\}$ and $\{G\}$ be related by the analogous expressions

$$\{\overset{(e)}{g}\} = [\overset{(e)}{\Lambda}]^T \{G\} \quad \text{or} \quad \{\overset{(e)}{g}\}^T = \{G\}^T [\overset{(e)}{\Lambda}] \tag{5.80}$$

Then, by using Equations (5.78), (5.79), and (5.80), the element integral $\overset{(e)}{I}$ of Equation (5.51), Section 5.6, may be written

$$I = (\tfrac{1}{2}) \{\overset{(e)}{T}\}^T [\overset{(e)}{k}] \{\overset{(e)}{T}\} + \{\overset{(e)}{g}\}^T \{\overset{(e)}{T}\}$$

$$= (\tfrac{1}{2}) \{T\}^T [\overset{(e)}{\Lambda}][\overset{(e)}{k}][\overset{(e)}{\Lambda}]^T \{T\} + \{G\}^T [\overset{(e)}{\Lambda}][\overset{(e)}{\Lambda}]^T \{T\} \tag{5.81}$$

Alternatively, $\overset{(e)}{I}$ may be written

$$\overset{(e)}{I} = (\tfrac{1}{2}) \{T\}^T [\overset{(e)}{K}] \{T\} - \{\overset{(e)}{F}\}^T \{T\} \tag{5.82}$$

where $[\overset{(e)}{K}]$ and $\{\overset{(e)}{F}\}^T$ are defined

$$[\overset{(e)}{K}] \overset{D}{=} [\overset{(e)}{\Lambda}] [\overset{(e)}{k}] [\overset{(e)}{\Lambda}]^T \tag{5.83}$$

and

$$\{\overset{(e)}{F}\}^T \overset{D}{=} - \{G\}^T [\overset{(e)}{\Lambda}] [\overset{(e)}{\Lambda}]^T \tag{5.84}$$

(The minus sign is introduced in the definition so that the governing finite element equations will have the same form as in the earlier chapters.) $[\overset{(e)}{K}]$ and $\{\overset{(e)}{F}\}$ are simply the element stiffness and forcing arrays in global dimensions.

Finally, by substituting from Equation (5.82) into Equation (5.45), in Section 5.6, the "global integral" I over R becomes

$$I = \sum_{e=1}^{N} \overset{(e)}{I} = \sum_{e=1}^{N} \langle (\tfrac{1}{2}) \{T\}^T \overset{(e)}{[K]} \{T\} - \overset{(e)}{\{F\}} \{T\} \rangle$$

$$= (\tfrac{1}{2}) \{T\}^T \left(\sum_{e=1}^{N} \overset{(e)}{[K]} \right) \{T\} - \left(\sum_{e=1}^{N} \overset{(e)}{\{F\}}^T \right) \{T\}$$

$$= (\tfrac{1}{2}) \{T\}^T [K] \{T\} - \{F\}^T \{T\} \tag{5.85}$$

where $[K]$ and $\{F\}$ are defined

$$[K] = \sum_{e=1}^{N} \overset{(e)}{[K]} \tag{5.86}$$

and

$$\{F\} = \sum_{e=1}^{N} \overset{(e)}{\{F\}} \tag{5.87}$$

$[K]$ and $\{F\}$ are, of course, the global stiffness and forcing arrays. The summations in Equations (5.86) and (5.87) together with the expansions of Equations (5.83) and (5.84) represent the assembly procedure.

5.9 GOVERNING EQUATIONS

In index form (that is, in terms of the nodal temperatures), Equation (5.85) might be written

$$I = (\tfrac{1}{2}) \sum_{i=1}^{n} \sum_{j=1}^{n} K_{ij} T_i T_j - \sum_{i=1}^{n} T_i F_i \tag{5.88}$$

The requirement of Equation (5.13) that $\delta I = 0$ is equivalent to requiring that

$$\partial I / \partial T_i = 0 \quad i = 1, \ldots, n \tag{5.89}$$

Substitution of Equation (5.88) into Equation (5.89) leads to the relations

$$\sum_{i=1}^{n} K_{ij} T_j = F_i \quad i = 1, \ldots, n \tag{5.90}$$

In matrix form, these relations may be written

$$[K]\{T\} = \{F\} \tag{5.91}$$

Equation (5.91) is readily recognized as our familiar set of governing equations. Hence, through the finite element modeling procedure the boundary value problem with the partial differential equation of Equations (5.1) to (5.3) has been converted into a system of algebraic equations. In the following section we will examine the development and solution of these equations for a specific example.

5.10 EXAMPLE: STEADY-STATE HEAT CONDUCTION IN A SQUARE REGION

As a simple example to illustrate some of the above procedures, consider the following problem: Let the temperature on three sides of a square region with unit sides be held at zero. Let the temperature on the fourth side be held at some constant value C other than zero. See Figure 5.9. The problem is then to determine the temperature distribution in the interior of the region.

Before solving this problem with the finite element procedure, let us first note that the boundary value problem may be stated as: Find $T(x, y)$ satisfying the partial differential equation

$$\partial^2 T/\partial x^2 + \partial^2 T/\partial y^2 = 0 \quad \begin{matrix} 0 \leqslant x \leqslant 1 \\ 0 \leqslant y \leqslant 1 \end{matrix} \tag{5.92}$$

with the boundary condition

$$T(0, y) = C, \quad T(1, y) = T(x, 0) = T(x, 1) = 0 \tag{5.93}$$

This problem may be solved using the separation of variables procedure leading to a Fourier series solution (see Problem 5.8).

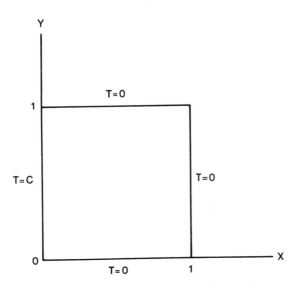

Figure 5.9   Square Region with Specified
Temperatures on the Boundary

Our procedure for obtaining a finite element modeling and solution to this problem might be outlined as follows: First, we will divide the region into elements, thus obtaining the finite element model. Next, we will develop the element stiffness matrices. Using these results, we will assemble the element stiffness matrices into the global stiffness matrix. This, in turn, will lead to the governing algebraic equations. Finally, we will solve these equations for the unknown temperatures in the interior of the region.

Therefore, to proceed with the solution, let us first divide the region into elements as shown in Figure 5.10. (As before, our model is kept simple to keep the analysis from becoming too detailed and thus obscuring the finite element procedure.) In Figure 5.10 we see that all elements have the same size and shape and that there are two orientations relative to the X-Y axes as shown in Figure 5.11. Let the local nodes for these two orientations, and hence types of elements, be as shown in Figure 5.11.

To determine the element stiffness matrixes $[k]^{(e)}$ ($e = 1, \ldots, N$), recall from Equation (5.52) that in general $[k]^{(e)}$ may be expressed as

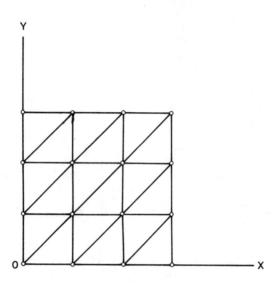

Figure 5.10 Finite Element Model of the Heat Conduction Region

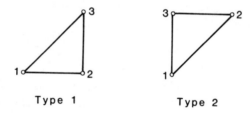

Type 1   Type 2

Figure 5.11 Element Orientations for the Finite Element Model

$$\overset{(e)}{[k]} = k\overset{(e)}{[B]}^T\overset{(e)}{[B]}\overset{(e)}{a} \qquad (5.94)$$

where k is the heat conduction coefficient, $\overset{(e)}{a}$ is the element area, and $\overset{(e)}{[B]}$ is the matrix obtained from the expressions

$$[B] = \begin{bmatrix} b_1 & b_2 & b_3 \\ c_1 & c_2 & c_3 \end{bmatrix}$$

$$= (\tfrac{1}{2}a) \begin{bmatrix} (y_2 - y_3) & (y_3 - y_1) & (y_1 - y_2) \\ (x_3 - x_2) & (x_1 - x_3) & (x_2 - x_1) \end{bmatrix} \qquad (5.95)$$

where (x_1, y_1), (x_2, y_2), and (x_3, y_3) are the nodal coordinates (see Problem 5.9).

From Figures 5.9 and 5.10, it is seen that the vertical and horizontal element sides all have a length of $1/3$. The element areas are thus $\overset{(e)}{a} = a = 1/18$. Consider the first type of orientation as defined in Figure 5.11: Since the horizontal and vertical sides have length $1/3$ the nodal coordinate differences in Equation (5.95) are

$$
\begin{bmatrix} (y_2 - y_3) & (y_3 - y_1) & (y_1 - y_2) \\ (x_3 - x_2) & (x_1 - x_3) & (x_2 - x_1) \end{bmatrix} = \begin{bmatrix} (-1/3) & (1/3) & 0 \\ 0 & (-1/3) & (1/3) \end{bmatrix}
$$

(5.96)

Hence, for a type 1 element, the matrix $[\overset{(e)}{B}]$ becomes

$$
\text{Type 1:} \quad [\overset{(e)}{B}] = \begin{bmatrix} -3 & 3 & 0 \\ 0 & -3 & 3 \end{bmatrix}
$$

(5.97)

Then, from Equation (5.94), the stiffness matrix $[\overset{(e)}{k}]$ for a type 1 element becomes

$$
\text{Type 1:} \quad [\overset{(e)}{k}] = \frac{k}{18} \begin{bmatrix} 9 & -9 & 0 \\ -9 & 18 & -9 \\ 0 & -9 & 9 \end{bmatrix}
$$

(5.98)

Similarly, for elements with type 2 orientation (see Figure 5.11), the stiffness matrix $[\overset{(e)}{k}]$ becomes

$$
\text{Type 2:} \quad [\overset{(e)}{k}] = \frac{k}{18} \begin{bmatrix} 9 & 0 & -9 \\ 0 & 9 & -9 \\ -9 & -9 & 18 \end{bmatrix}
$$

(5.99)

To assemble these element stiffness matrices into the global stiffness matrix $[K]$, let the elements and nodes of the model of Figure 5.10 be numbered as shown in Figure 5.12. Thus there are 18 elements (9 of each type as defined in Figure 5.11) and there are 16 nodes. The incidence matrices $[\Lambda]^{(e)}$ introduced in Section 5.8 can now be determined as follows: First, the matching, or identification, between the local and global nodes for each element can be recorded as in Table 5.1. Next, note that since there are 16 nodes there are 16 global temperatures. Hence, the incidence matrices have 3 rows and 16 columns. The nonzero entries in the incidence matrices are obtained by placing 1's in the columns corresponding to the global node numbers of Table 5.1. For example, for element (1) the incidence matrix is

$$[\Lambda]^{(1)} = \begin{bmatrix} 1 & 0 & 0 & 0 & 0 & 0 & 0 & 0 & 0 & 0 & 0 & 0 & 0 & 0 & 0 & 0 \\ 0 & 1 & 0 & 0 & 0 & 0 & 0 & 0 & 0 & 0 & 0 & 0 & 0 & 0 & 0 & 0 \\ 0 & 0 & 0 & 0 & 0 & 1 & 0 & 0 & 0 & 0 & 0 & 0 & 0 & 0 & 0 & 0 \end{bmatrix}$$

$$(5.96)$$

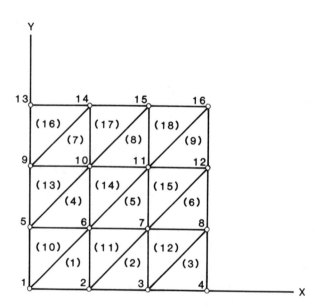

Figure 5.12 Node and Element Numbering for the Model of Figure 5.10

Table 5.1 Matching Between Element and Global Nodes

Element Nodes	Elements (1)	(2)	(3)	(4)	(5)	(6)	(7)	(8)	(9)	(10)	(11)	(12)	(13)	(14)	(15)	(16)	(17)	(18)
1	1	2	3	5	6	7	9	10	11	1	2	3	5	6	7	9	10	11
2	2	3	4	6	7	8	10	11	12	6	7	8	10	11	12	14	15	16
3	6	7	8	10	11	12	14	15	16	5	6	7	9	10	11	13	14	15

Global Nodes

For element (12), the incidence matrix is

$$
\overset{(12)}{[\Lambda]} = \begin{bmatrix} 0 & 0 & 1 & 0 & 0 & 0 & 0 & 0 & 0 & 0 & 0 & 0 & 0 & 0 & 0 & 0 \\ 0 & 0 & 0 & 0 & 0 & 0 & 0 & 1 & 0 & 0 & 0 & 0 & 0 & 0 & 0 & 0 \\ 0 & 0 & 0 & 0 & 0 & 0 & 1 & 0 & 0 & 0 & 0 & 0 & 0 & 0 & 0 & 0 \end{bmatrix}
$$

$$(5.97)$$

Using the incidence matrices and the element stiffness matrices, the element stiffness matrices $\overset{(e)}{[K]}$ in global dimensions may be obtained [see Section 5.8, Equation (5.83)]:

$$
\overset{(e)}{[K]} = \overset{(e)}{[\Lambda]} \overset{(e)}{[k]} \overset{(e)}{[\Lambda]}^T
$$

$$(5.98)$$

Then, by addition over the elements, the global stiffness matrix $[K]$ is obtained [see Section 5.8, Equation (5.86)]

$$
[K] = \sum_{e=1}^{18} \overset{(e)}{[K]}
$$

$$(5.99)$$

However, we can avoid some of the details of this procedure by simply noting that the entries in the rows and columns of Table 5.1 designate the rows and columns of the global stiffness matrix. For example, for element (1), the components of the element stiffness matrix are inserted into the global stiffness matrix as follows:

$$
\overset{(1)}{k_{11}} \rightarrow K_{11} \quad \overset{(1)}{k_{12}} \rightarrow K_{12} \quad \overset{(1)}{k_{13}} \rightarrow K_{16}
$$

$$
\overset{(1)}{k_{21}} \rightarrow K_{21} \quad \overset{(1)}{k_{22}} \rightarrow K_{22} \quad \overset{(1)}{k_{23}} \rightarrow K_{26}
$$

$$(5.100)$$

$$
\overset{(1)}{k_{31}} \rightarrow K_{61} \quad \overset{(1)}{k_{32}} \rightarrow K_{62} \quad \overset{(1)}{k_{33}} \rightarrow K_{66}
$$

Similarly, for say elements (12) and (18) we have the assignments

$$
\overset{(12)}{k_{11}} \to K_{33} \qquad \overset{(12)}{k_{12}} \to K_{38} \qquad \overset{(12)}{k_{13}} \to K_{37}
$$

$$
\overset{(12)}{k_{21}} \to K_{83} \qquad \overset{(12)}{k_{22}} \to K_{88} \qquad \overset{(12)}{k_{23}} \to K_{87} \tag{5.101}
$$

$$
\overset{(12)}{k_{31}} \to K_{73} \qquad \overset{(12)}{k_{23}} \to K_{78} \qquad \overset{(12)}{k_{33}} \to K_{77}
$$

and

$$
\overset{(18)}{k_{11}} \to K_{11,11} \qquad \overset{(18)}{k_{12}} \to K_{11,16} \qquad \overset{(18)}{k_{13}} \to K_{11,15}
$$

$$
\overset{(18)}{k_{21}} \to K_{16,11} \qquad \overset{(18)}{k_{22}} \to K_{16,16} \qquad \overset{(18)}{k_{23}} \to K_{16,15} \tag{5.102}
$$

$$
\overset{(18)}{k_{31}} \to K_{15,11} \qquad \overset{(18)}{k_{32}} \to K_{15,16} \qquad \overset{(18)}{k_{33}} \to K_{15,15}
$$

5.11   SOLUTION TO THE EXAMPLE HEAT CONDUCTION PROBLEM

By knowing the global stiffness matrix of Equation (5.99), together with the given boundary conditions, we are in an excellent position to obtain the temperature distribution in the interior of the region—that is, at the interior nodes. Specifically, from the given boundary conditions of Equation (5.93) with $C = 10$, we have

$$
T(0, y) = 10, \quad T(1, y) = T(x, 0) = T(x, 1) = 0 \tag{5.103}
$$

From Figure 5.12, this means that

$$
T_1 = T_{13} = 5 \quad \text{and} \quad T_5 = T_9 = 10 \tag{5.104}
$$

and that

$$
T_2 = T_3 = T_4 = T_8 = T_{12} = T_{14} = T_{15} = T_{16} = 0 \tag{5.105}
$$

Hence, the unknown nodal temperature are T_6, T_7, T_{10}, and T_{11}. We can obtain expression governing these temperatures from the governing equations (5.90), by considering the equations corresponding to $i = 6, 7, 10$, and 11 (see Problem 5.12). This leads to

$$K_{61}(5) + K_{65}(10) + K_{66}T_6 + K_{67}T_7 + K_{69}(10)$$
$$+ K_{6,10}T_{10} + K_{6,11}T_{11} + K_{6,13}(5) = 0 \qquad (5.106)$$

$$K_{71}(5) + K_{75}(10) + K_{76}T_6 + K_{77}T_7 + K_{79}(10)$$
$$+ K_{7,10}T_{10} + K_{7,11}T_{11} + K_{7,13}(5) = 0 \qquad (5.107)$$

$$K_{10,1}(5) + K_{10,5}(10) + K_{10,6}T_6 + K_{10,7}T_7 + K_{10,9}(10)$$
$$+ K_{10,10}T_{10} + K_{10,11}T_{11} + K_{10,13}(5) = 0 \qquad (5.108)$$

and

$$K_{11,1}(5) + K_{11,5}(10) + K_{11,6}T_6 + K_{11,7}T_7 + K_{11,9}(10)$$
$$+ K_{11,10}T_{10} + K_{11,11}T_{11} + K_{11,13}(5) = 0 \qquad (5.109)$$

where the terms with zero temperature have been omitted.

Using Equations (5.98) and (5.99) together with the procedure exemplified by Equations (5.100), (5.101), and (5.102), we find the following expressions for the coefficients in the above four equations

$$K_{61} = \overset{(1)}{k_{31}} + \overset{(10)}{k_{21}} = 0$$

$$K_{65} = \overset{(4)}{k_{21}} + \overset{(10)}{k_{23}} = -1$$

$$K_{67} = \overset{(5)}{k_{12}} + \overset{(11)}{k_{32}} = -1 \qquad (5.110)$$

$$K_{66} = \overset{(1)}{k_{33}} + \overset{(4)}{k_{22}} + \overset{(5)}{k_{11}} + \overset{(10)}{k_{22}} + \overset{(11)}{k_{33}} + \overset{(14)}{k_{11}} = 4$$

$$K_{6,10} = \overset{(4)}{k_{23}} + \overset{(14)}{k_{13}} = -1$$

$$K_{6,11} = \overset{(5)}{k_{13}} + \overset{(14)}{k_{12}} = 0 \qquad (5.110)$$

$$K_{69} = K_{6,13} = 0$$

$K_{76} = K_{67} = -1$

$K_{77} = \overset{(2)}{k_{33}} + \overset{(5)}{k_{22}} + \overset{(6)}{k_{11}} + \overset{(11)}{k_{22}} + \overset{(12)}{k_{33}} + \overset{(15)}{k_{11}} = 4$

$K_{7,11} = \overset{(5)}{k_{23}} + \overset{(15)}{k_{13}} = -1$ (5.111)

$K_{71} = K_{75} = K_{79} = K_{7,10} = K_{7,13} = 0$

$K_{10,5} = \overset{(4)}{k_{31}} + \overset{(13)}{k_{12}} = 0$

$K_{10,6} = K_{6,10} = -1$

$K_{10,9} = \overset{(7)}{k_{21}} + \overset{(13)}{k_{23}} = -1$

$K_{10,10} = \overset{(4)}{k_{33}} + \overset{(7)}{k_{22}} + \overset{(8)}{k_{11}} + \overset{(13)}{k_{22}} + \overset{(14)}{k_{33}} + \overset{(17)}{k_{11}} = 4$ (5.112)

$K_{10,11} = \overset{(8)}{k_{12}} + \overset{(14)}{k_{23}} = -1$

$K_{10,7} = K_{7,10} = 0$

$K_{10,1} = K_{10,13} = 0$

$K_{11,6} = K_{6,11} = 0$

$K_{11,7} = K_{7,11} = -1$

$K_{11,10} = K_{10,11} = -1$ (5.113)

$K_{11,11} = \overset{(5)}{k_{33}} + \overset{(8)}{k_{22}} + \overset{(9)}{k_{11}} + \overset{(14)}{k_{22}} + \overset{(15)}{k_{33}} + \overset{(18)}{k_{11}} = 4$

$K_{11,1} = K_{11,5} = K_{11,9} = K_{11,13} = 0$

Hence, by substituting these results, Equations (5.106) to (5.109) become

$4T_6 - T_7 - T_{10} = 10$ (5.114)

$$-T_6 + 4T_7 - T_{11} = 0 \qquad\qquad (5.115)$$

$$-T_6 + 4T_{10} - T_{11} = 10 \qquad\qquad (5.116)$$

and

$$-T_7 - T_{10} + 4T_{11} = 0 \qquad\qquad (5.117)$$

By solving these equations for T_6, T_7, T_{10}, and T_{11}, we obtain

$$T_6 = T_{10} = 15/4 = 3.75 \qquad\qquad (5.118)$$

and

$$T_7 = T_{11} = 5/4 = 1.25 \qquad\qquad (5.119)$$

By comparing the above results with those of the exact solution of Problem 5.9, we find a remarkable agreement even for this relatively coarse finite element model.

5.12 INTRODUCTION TO PLANE STRESS/PLANE STRAIN

The finite element method has been used extensively with plane elasticity problems. Before we discuss the specific procedures used to solve these problems, however, it might be helpful to review the fundamental concepts of elasticity theory and, specifically, of plane stress and plane strain elasticity. (The reader who is unfamiliar with this subject may want to refer to a text on advanced strength of materials or elasticity theory.*)

Let the displacement at a point of an elastic body be denoted by the components u, v, and w relative to a cartesian coordinate system XYZ. The *normal* strains ϵ_x, ϵ_y, and ϵ_z may then be expressed as

$$\epsilon_x = \partial u/\partial x, \quad \epsilon_y = \partial v/\partial y, \quad \text{and} \quad \epsilon_z = \partial w/\partial z \qquad (5.120)$$

---

*See a brief list of references at the end of this section.

Correspondingly, the shear strains γ_{xy}, γ_{yz}, and γ_{zx} may be expressed as

$$\gamma_{xy} = \partial u/\partial x + \partial u/\partial y, \quad \gamma_{yz} = \partial w/\partial y + \partial v/\partial z, \quad \text{and}$$
$$\gamma_{zx} = \partial u/\partial z + \partial w/\partial x \qquad (5.121)$$

[The reader may recall that shear strain components are often defined with a factor of ½ multiplying the terms on the right side of Equation (5.121). If this is done, the normal strains together with the shear strains form the components of a strain *tensor*. For our purposes here however, it is computationally convenient to delete the ½ in the definition.]

The normal strains are related to the normal stresses σ_x, σ_y, and σ_z by the familiar linear relations

$$\epsilon_x = (1/E)\,[\sigma_x - \nu(\sigma_y + \sigma_z)]$$
$$\epsilon_y = (1/E)\,[\sigma_y - \nu(\sigma_z + \sigma_x)]$$

(5.122)

and

$$\epsilon_x = (1/E)\,[\sigma_z - \nu(\sigma_x + \sigma_y)]$$

where E is Young's modulus of elasticity and ν is Poisson's ratio. Similarly, the shear strains, as we have defined them, are related to the shear stresses τ_{xy}, τ_{yz} and τ_{zx} by the simple expressions

$$\gamma_{xy} = (1/G)\tau_{xy}, \quad \gamma_{yz} = (1/G)\tau_{yz}, \quad \text{and} \quad \gamma_{zx} = (1/G)\tau_{zx}$$

(5.123)

where G is the shear modulus or "modulus of rigidity." G may be expressed in terms of E and ν as

$$G = E/2(1 + \nu) \qquad (5.124)$$

Examining the "force" equilibrium of the elastic body leads to the *equilibrium* equations

$$\partial\sigma_x/\partial x + \partial\tau_{xy}/\partial y + \partial\tau_{xz}/\partial z + F_x = 0$$

$$\partial\tau_{yx}/\partial x + \partial\sigma_y/\partial y + \partial\tau_{yz}/\partial z + F_y = 0 \qquad (5.125)$$

and

$$\partial\tau_{zx}/\partial x + \partial\tau_{zy}/\partial y + \partial\sigma_z/\partial z + F_z = 0$$

where F_x, F_y, and F_z are the components of an applied body force (per unit volume). Similarly, examining the "moment" equilibrium of the body leads to the simple relations

$$\tau_{xy} = \tau_{yx}, \quad \tau_{yz} = \tau_{zx}, \quad \text{and} \quad \tau_{zx} = \tau_{xz} \qquad (5.126)$$

[Note that Equations (5.126) are consistent with the definitions of the shear strains in Equations (5.121) together with the stress strain relations of Equations (5.123). That is, both the shear stress and the shear strain are "symmetric" in that the subscripts may be interchanged.]

The "displacement formulation" of boundary value problems in elasticity is obtained by solving Equations (5.122) and (5.123) for the stresses in terms of the strains and then by substituting into the equilibrium equations (5.125). [Note that the equilibrium equations (5.126) are identically satisfied.] Then by substituting from Equations (5.120) and (5.121) we obtain the following three equations for the three unknown displacement components u, v, and w:

$$(\lambda + G)(\partial^2 u/\partial x^2 + \partial^2 v/\partial y \partial x + \partial^2 w/\partial z \partial x)$$
$$+ G(\partial^2 u/\partial x^2 + \partial^2 u/\partial y^2 + \partial^2 u/\partial z^2) + F_x = 0 \qquad (5.127)$$

$$(\lambda + G)(\partial^2 u/\partial x \partial y + \partial^2 v/\partial y^2 + \partial^2 w/\partial z \partial x)$$
$$+ G(\partial^2 v/\partial x^2 + \partial^2 v/\partial y^2 + \partial^2 v/\partial z^2) + F_y = 0 \qquad (5.128)$$

and

$$(\lambda + G)(\partial^2 u/\partial x \partial z + \partial^2 v/\partial y \partial z + \partial^2 w/\partial z^2)$$
$$+ G(\partial^2 w/\partial x^2 + \partial^2 w/\partial y^2 + \partial^2 w/\partial z^2) + F_z = 0 \qquad (5.129)$$

where λ is

$$\lambda = Ev/(1 + v)(1 - 2v) \qquad (5.130)$$

When Equations (5.127), (5.128), and (5.129) are solved subject to suitable boundary conditions, the strains are obtained through Equations (1.120) and (5.121) and then the stresses may be obtained through Equations (5.122) and (5.123).

A variational formulation of boundary value problems in elasticity may be obtained through the "principle of Minimum Total Potential Energy" as follows: Let an elastic body occupy a region R bounded by a surface S. Next, introduce displacement and loading arrays defined

$$\{u\} = \begin{bmatrix} u \\ v \\ w \end{bmatrix} \qquad \{F\} = \begin{bmatrix} F_x \\ F_y \\ F_z \end{bmatrix} \qquad \{T\} = \begin{bmatrix} T_x \\ T_y \\ T_z \end{bmatrix} \qquad (5.131)$$

where $\{T\}$ is sometimes called the "traction array." It consists of the X, Y, and Z components of a force per unit area T, applied on the surface S. Finally, let the "strain energy density" W be defined

$$W = (\tfrac{1}{2})(\epsilon_x \sigma_x + \epsilon_y \sigma_y + \epsilon_z \sigma_z + \gamma_{xy} \tau_{xy} + \gamma_{yz} \tau_{yz} + \gamma_{zx} \tau_{zx})$$
$$(5.132)$$

Then, the principle of Minimum Total Potential Energy states that: The displacement components u, v, and w which satisfy the boundary value problem outlined in Equations (5.127), (5.128), and (5.129), together with suitable boundary conditions, also results in a minimum value of the potential P where P is given by the expression

$$P = \int_R W dR - \int_R \{u\}^T \{F\} dR - \int_S \{u\}^T \{T\} dS \qquad (5.133)$$

This principle, which is directly analogous to the variational principle developed in Section 5.1 for the heat conduction problem, will enable us to conveniently develop the finite element formulation of the elasticity boundary value problem.

The *plane stress* approximation which is useful with thin flat elastic bodies is developed by taking the stresses in the direction normal to the midplane of the body to be zero. That is, if the Z axis is normal to the plane of the body, then

$$\sigma_z = \tau_{xz} = \tau_{yz} = 0 \qquad (5.134)$$

The stress-strain equations (5.122) and (5.123) then become

$$\epsilon_x = (1/E)(\sigma_x - \nu\sigma_y) \tag{5.135}$$

$$\epsilon_y = (1/E)(\sigma_y - \nu\sigma_x) \tag{5.136}$$

$$\epsilon_x = (-\nu/E)(\sigma_x + \sigma_y) \tag{5.137}$$

and

$$\gamma_{xy} = (1/G)\tau_{xy} \tag{5.138}$$

Next, let the stress and strain arrays $\{\sigma\}$ and $\{\epsilon\}$, in the plane of the body be defined

$$\{\sigma\} = \begin{bmatrix} \sigma_x \\ \sigma_y \\ \sigma_{xy} \end{bmatrix} \quad \text{and} \quad \{\epsilon\} = \begin{bmatrix} \epsilon_x \\ \epsilon_y \\ \gamma_{xy} \end{bmatrix} \tag{5.139}$$

Then, by using Equations (5.135), (5.136), and (5.137), $\{\sigma\}$ and $\{\epsilon\}$ may be related by the expressions

$$\{\sigma\} = [E]\{\epsilon\} \quad \text{and} \quad \{\epsilon\} = [E]^{-1}\{\sigma\} \tag{5.140}$$

where the matrix $[E]$ is

$$[E] = \frac{E}{(1 - \nu^2)} \begin{bmatrix} 1 & \nu & 0 \\ \nu & 1 & 0 \\ 0 & 0 & (1 - \nu)/2 \end{bmatrix} \tag{5.141}$$

From Equations (5.132) and (5.134), it is seen that the strain energy density may be expressed in terms of $\{\epsilon\}$, $\{\sigma\}$, and $[E]$ as

$$W = (½) \{\epsilon\}^T \{\sigma\} = (½) \{\epsilon\}^T [E] \{\epsilon\} \tag{5.142}$$

Finally, for the plane stress approximation, the potential P of Equation (5.133) takes the form

$$P = \int_R (\tfrac{1}{2}) \{\epsilon\}^T [E] \{\epsilon\} \, dR - \int_R \{u\}^T \{F\} \, dR - \int_C \{u\}^T \{T\} \, dC \tag{5.143}$$

where C is the boundary curve of the plane region R and where $\{u\}$, $\{F\}$, and $\{T\}$ are the two-dimensional arrays

$$\{u\} = \begin{bmatrix} u \\ v \end{bmatrix}, \quad \{F\} = \begin{bmatrix} F_x \\ F_y \end{bmatrix}, \quad \text{and} \quad \{T\} = \begin{bmatrix} T_x \\ T_y \end{bmatrix} \tag{5.144}$$

The *plane strain* approximation which is useful with long cylindrical bodies is developed by taking the displacement in the direction of the cylinder axis to be zero and the derivatives in the direction of the cylinder axis to be zero. If the Z axis is along the cylinder axis, these assumptions lead to the expressions

$$\omega = 0 \quad \text{and} \quad \partial(\ )/\partial z = 0 \tag{5.145}$$

From Equations (5.120) and (5.121) these expressions mean that

$$\epsilon_z = \gamma_{zx} = \gamma_{zy} = 0 \tag{5.146}$$

Hence, for plane strain, the stress-strain equations (5.122) and (5.123) become (see Problem 5.15)

$$\epsilon_x = [(1 + \nu)/E] [(1 - \nu)\sigma_x - \nu\sigma_y] \tag{5.147}$$

$$\epsilon_y = [(1 + \nu)/E] [(1 - \nu)\sigma_y - \nu\sigma_x] \tag{5.148}$$

and

$$\gamma_{xy} = (1/G)\tau_{xy} \tag{5.149}$$

As with plane stress, let the stress and strain arrays $\{\sigma\}$ and $\{\epsilon\}$ be defined in the plane normal to the cylinder axis, with the same expressions as in Equations (5.139). Then for plane strain $\{\sigma\}$ and $\{\epsilon\}$ are related by the expression

$$\{\sigma\} = [\hat{E}] \{\epsilon\} \quad \text{and} \quad \{\epsilon\} = [\hat{E}]^{-1} \{\sigma\} \qquad (5.150)$$

where $[\hat{E}]$ is given by (see Problem 5.16)

$$\hat{E} = \frac{E}{(1+\nu)(1-2\nu)} \begin{bmatrix} (1-\nu) & \nu & 0 \\ \nu & (1-\nu) & 0 \\ 0 & 0 & (1-2\nu)/2 \end{bmatrix} \qquad (5.151)$$

Finally, for plane strain, the strain energy density w and the potential P have the same forms as in Equations (5.142) and (5.143) except that $[E]$ is replaced by $[\hat{E}]$.

In the following section we will outline the finite element procedure for minimizing P. This procedure will be directly analogous to the procedure used in the foregoing section with heat conduction. There is, however, one additional relation which will be helpful in that formulation: Let the differential operator $[\partial]$ be defined as

$$[\partial] = \begin{bmatrix} \partial/\partial x & 0 \\ 0 & \partial/\partial y \\ \partial/\partial x & \partial/\partial y \end{bmatrix} \qquad (5.152)$$

Then it is readily seen that the two dimensional array $\{u\}$ and the plane array $\{\epsilon\}$ are related by

$$\{\epsilon\} = [\partial] \{u\} \qquad (5.153)$$

5.13 FINITE ELEMENT FORMULATION OF THE PLANE STRESS/PLANE STRAIN BOUNDARY VALUE PROBLEM

As mentioned in the foregoing section, our development of a finite element formulation of the plane stress/plane strain boundary value problem will closely follow that already established in the earlier sections for the heat conduction problem. Hence, we will not emphasize all the details of the development, but instead we will focus our attention on those features which are distinctive to plane stress and plane strain problems.

The variational statement of the plane stress/plane strain boundary value problem might be stated simply as: Find the displacement array $\{u\}$ such that $\delta P = 0$ where P is the potential of Equation (5.143) and $\{u\}$ is the two-dimensional array

$$\{u\} = \begin{bmatrix} u \\ v \end{bmatrix} \tag{5.154}$$

Therefore, our objective will be to determine $\{u\}$ throughout a given region R. As before, let us divide the region R into N elements containing a total of n nodes as shown for example in Figure 5.13. Next, on a typical element e let us introduce as before, linear interpolation or shape functions $\overset{(e)}{N_i}(x, y)$ ($i = 1, 2, 3$) having the value: 1 at local node i and the value: 0 at the other two nodes of the element. Also, as before, let $\overset{(e)}{N_i}$ be zero at all points of R except those of (e). Finally, let the element displacement array on (e) be denoted by $\overset{(e)}{\{u\}}$ and let the components $\overset{(e)}{u}$ and $\overset{(e)}{v}$ be expressed as

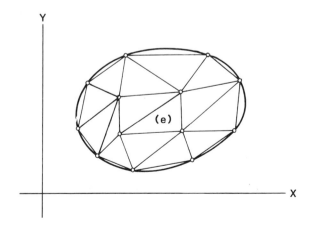

Figure 5.13   Finite Element Division of Region R

$$\overset{(e)}{u} = \overset{(e)}{u_1}\overset{(e)}{N_1} + \overset{(e)}{u_2}\overset{(e)}{N_2} + \overset{(e)}{u_3}\overset{(e)}{N_3} \tag{5.155}$$

and

$$\overset{(e)}{v} = \overset{(e)}{v_1}\overset{(e)}{N_1} + \overset{(e)}{v_2}\overset{(e)}{N_2} + \overset{(e)}{v_3}\overset{(e)}{N_3} \tag{5.156}$$

At this point it is convenient to introduce a second element displacement array $\{\overset{(e)}{d}\}$ containing the six local or element nodal displacements as

$$\{\overset{(e)}{d}\}^T = [\overset{(e)}{u_1}, \overset{(e)}{v_1}, \overset{(e)}{u_2}, \overset{(e)}{v_2}, \overset{(e)}{u_3}, \overset{(e)}{v_3}] \tag{5.157}$$

Then $\{\overset{(e)}{u}\}$ and $\{\overset{(e)}{d}\}$ are related by the expression

$$\{\overset{(e)}{u}\} = [\overset{(e)}{N}] \{\overset{(e)}{d}\} \tag{5.158}$$

where $[N]$ is the 2 × 6 array defined by

$$[\overset{(e)}{N}] = \begin{bmatrix} \overset{(e)}{N_1} & 0 & \overset{(e)}{N_2} & 0 & \overset{(e)}{N_3} & 0 \\ 0 & \overset{(e)}{N_1} & 0 & \overset{(e)}{N_2} & 0 & \overset{(e)}{N_3} \end{bmatrix} \tag{5.159}$$

From Equation (5.153), let the element strain array $\{\overset{(e)}{\epsilon}\}$ be expressed as

$$\{\overset{(e)}{\epsilon}\} = \begin{bmatrix} \overset{(e)}{\epsilon_x} \\ \overset{(e)}{\epsilon_y} \\ \overset{(e)}{\gamma_{xy}} \end{bmatrix} = [\partial] \{\overset{(e)}{u}\} = [\partial] [N] \{\overset{(e)}{d}\} \overset{D}{=} [\overset{(e)}{B}] \{\overset{(e)}{d}\} \tag{5.160}$$

where $[\overset{(e)}{B}]$ is defined here as $[\partial][\overset{(e)}{N}]$. Recall from Equation (5.28), Section 5.4 that $\{\overset{(e)}{N}\}^T$ may be expressed as

$$\{ \overset{(e)}{N} \}^T = [\overset{(e)}{N_1}, \overset{(e)}{N_2}, \overset{(e)}{N_3}] = [1 \ x \ y] \ [A]^{-1}$$

$$= [1 \ x \ y] \begin{bmatrix} a_1 & a_2 & a_3 \\ b_1 & b_2 & b_3 \\ c_1 & c_2 & c_3 \end{bmatrix} \qquad (5.161)$$

where from Equation (5.31) the array $[A]^{-1}$ is

$$[A]^{-1} = \begin{bmatrix} a_1 & a_2 & a_3 \\ b_1 & b_2 & b_3 \\ c_1 & c_2 & c_3 \end{bmatrix}$$

$$= \frac{1}{2a} \begin{bmatrix} (x_2 y_3 - x_3 y_2) & (x_3 y_1 - x_1 y_3) & (x_1 y_2 - x_2 y_1) \\ (y_2 - y_3) & (y_3 - y_1) & (y_1 - y_2) \\ (x_3 - x_1) & (x_1 - x_2) & (x_2 - x_3) \end{bmatrix}$$

$$(5.162)$$

where a is the traingle area and $(\overset{(e)}{x_i}, \overset{(e)}{y_i})$ $(i = 1, 2, 3)$ are the coordinates of the local node i. Hence, by substituting from Equation (5.161) and by using Equation (5.152) the array $[\overset{(e)}{B}]$ is found to be

$$\overset{(e)}{[B]} = \begin{bmatrix} b_1 & 0 & b_2 & 0 & b_3 & 0 \\ 0 & c_1 & 0 & c_2 & 0 & c_3 \\ c_1 & b_1 & c_2 & b_2 & c_3 & c_3 \end{bmatrix} \qquad (5.163)$$

Finally, let $\{ D \}$ be the array of global displacement components defined by the expression

$$\{ D \}^T = [U_1 \ V_1 \ U_2 \ V_2 \cdots U_3 \ V_3] \qquad (5.164)$$

where (U_I, V_I) are the displacement components of node I.

Then, by using a table similar to that of Table 5.1 of Section 5.10, we can develop element incidence matrices $[\overset{(e)}{\Lambda}]$ such that

$$\{\overset{(e)}{d}\} = [\overset{(e)}{\Lambda}]\{D\} \tag{5.165}$$

5.14 GOVERNING EQUATIONS FOR THE PLANE STRESS/PLANE STRAIN BOUNDARY VALUE PROBLEM

We are now in a position to develop the algebraic finite element equations which will provide the approximate solution to the plane stress/plane strain boundary value problem. First, let us note that since the interpolation functions $\{\overset{(e)}{N}\}$ are zero except on element (e), the potential function P may be expressed as

$$P = \sum_{e=1}^{N} \overset{(e)}{P} \tag{5.166}$$

where $\overset{(e)}{P}$ is

$$\overset{(e)}{P} = (\tfrac{1}{2}) \int_{(e)} \{\overset{(e)}{\epsilon}\}^T [E]\{\overset{(e)}{\epsilon}\} \, de - \int_{(e)} \{\overset{(e)}{u}\}^T \{F\} \, d(e)$$
$$- \int_{s} \{\overset{(e)}{u}\}^T \{T\} \, ds \tag{5.167}$$

where s is the boundary of (e). By substituting from Equations (5.158) and (5.160) for $\{\overset{(e)}{u}\}$ and $\{\overset{(e)}{\epsilon}\}$, $\overset{(e)}{P}$ may be written

$$\overset{(e)}{P} = (\tfrac{1}{2}) \{\overset{(e)}{d}\}^T \left\{ \int_{(e)} [B]^T [E] [B] \, d(e) \right\} \{\overset{(e)}{d}\}$$

$$- \{d\}^T \int_{(e)} [\overset{(e)}{N}]^T \{F\} \, d(e) - \{\overset{(e)}{d}\}^T \int_{s} [\overset{(e)}{N}]^T \{T\} \, ds \tag{5.168}$$

or

$$\overset{(e)}{P} = (\tfrac{1}{2}) \{\overset{(e)}{d}\}^T [\overset{(e)}{k}] \{\overset{(e)}{d}\} - \{\overset{(e)}{d}\}^T \{\overset{(e)}{f}\} - \{\overset{(e)}{d}\}^T \{\overset{(e)}{t}\} \tag{5.169}$$

where $[\overset{(e)}{k}]$, $\{\overset{(e)}{f}\}$, and $\{\overset{(e)}{t}\}$ are the element "stiffness" and "forcing" arrays given by

$$\overset{(e)}{[k]} = \int\limits_{(e)} [B]^T [E] [B] \, d(e) \tag{5.170}$$

$$\{\overset{(e)}{f}\} = \int\limits_{(e)} \overset{(e)}{[N]}^T \{F\} \, d(e) = \int\limits_{(e)} \overset{(e)}{[N]}^T \overset{(e)}{[N]} \, d(e) \{\overset{(e)}{g}\} \tag{5.171}$$

and

$$\{\overset{(e)}{t}\} = \int\limits_{s} \overset{(e)}{[N]}^T \{T\} \, ds = \int\limits_{s} \overset{(e)}{[N]}^T \overset{(e)}{[N]} \, ds \, \{\overset{(e)}{\tau}\} \tag{5.172}$$

where $\{g\}$ and $\{\tau\}$ are the nodal values of $\{F\}$ and $\{T\}$.

By using the incidence matrices as introduced in Equation (5.153), $\overset{(e)}{P}$ may be further developed in terms of the global displacement array $\{D\}$ as

$$\overset{(e)}{P} = (\tfrac{1}{2}) \{D\}^T \overset{(e)}{[\Lambda]}^T \overset{(e)}{[k]} \overset{(e)}{[\Lambda]} \{D\} - \{D\}^T \overset{(e)}{[\Lambda]}^T \{\overset{(e)}{f}\}$$

$$- \{D\}^T \overset{(e)}{[\Lambda]}^T \{\overset{(e)}{t}\} \tag{5.173}$$

or as

$$\overset{(e)}{P} = (\tfrac{1}{2}) \{D\}^T \overset{(e)}{[K]} \{D\} - \{D\}^T \{\overset{(e)}{F}\} - \{D\}^T \{\overset{(e)}{T}\} \tag{5.174}$$

where $[\overset{(e)}{K}]$, $\{\overset{(e)}{F}\}$, and $\{\overset{(e)}{T}\}$ are the element stiffness and forcing arrays in global dimensions, and are defined by inspection and comparison of Equations (5.173) or (5.174).

Finally, by substituting Equation (5.174) into (5.166), the potential P may be written

$$P = (\tfrac{1}{2}) \{D\}^T [K] \{D\} - \{D\}^T \{F\} \tag{5.175}$$

where the global stiffness matrix $[K]$ and the global forcing function $\{F\}$ are

$$[K] = \sum_{e=1}^{N} \overset{(e)}{[K]} \tag{5.176}$$

and

$$\{F\} = \sum_{e=1}^{N} (\{\overset{(e)}{F}\} + \{\overset{(e)}{T}\})$$ (5.177)

Then, by setting the derivative of P with respect to the components of $\{D\}$ equal to zero (to minimize P), we obtain the now very familiar expression

$$[K]\{D\} = \{F\}$$ (5.178)

5.15 CLOSURE

The foregoing discussion on the finite element formulation of the plane stress/plane strain boundary value problem illustrates the basic approach generally taken in the finite element formulation of boundary value problems in general. That is, the basic steps outlined above form the fundamental procedure of the finite element method. First, the variational statement of the problem is obtained. This statement may be obtained from either a physical principle such as the minimum total potential as in elasticity theory, or by integrating by parts using a variational operator as was done for the heat conduction problem. Next, the region of interest needs to be divided into elements. A judicious choice of the number, size, and shape of the elements can have a marked effect upon the accuracy and efficiency of the subsequent calculations. Also, the local/global numbering of the elements forms the basis of the assembly procedure. Following this, element interpolation functions or shape functions need to be selected to express the dependent variable (or variables) in terms of the independent variables. This is a means of separating the nodal values (and hence, the dependent variables) from the independent variables. Then, the introduction of the incidence matrices and the substitution into the variational equations generates the assembly procedure. Finally, by satisfying the variational statement of the problem (that is, by minimizing the integral), the governing finite element algebraic equations are obtained. The solution of these equations then depends upon the given boundary conditions.

PROBLEMS

Section 5.2

5.1 Using the procedures of Section 5.2, develop a variational formulation of Equation (5.1), (5.2), and (5.3) with $k = k(x, y)$ and $h \neq 0$.

Answer: $\delta I = 0$

where

$$I = (\tfrac{1}{2}) \int_R [k(\nabla T)^2 + 2fT]\, dR - \int_{C_2} Tkhd C_2$$

Section 5.4

5.2 Show that $\Delta = 2a$ [see Equation (5.30)].

Suggestion: Introduce vectors p and q connecting the nodes 1 and 3 and the nodes 1 and 2 as shown in Figure P5.2. Then recall that the triangular area is $(\tfrac{1}{2})|p|\,|q|\sin\theta = (\tfrac{1}{2})|p \times q|$ where θ is the angle between p and q.

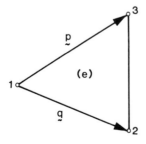

Figure P5.2   Vectors Along the Element Edges

5.3 By using the results of Problem 5.2 and in view of Equation (5.30), show that

$$a = (\tfrac{1}{2}) \begin{bmatrix} 1 & x_1 & y_1 \\ 1 & x_2 & y_2 \\ 1 & x_3 & y_3 \end{bmatrix}$$

5.4 Show that $N_1 + N_2 + N_3 = 1$.

Section 5.7

5.5 Consider the triangular element (e) positioned relative to the coordinate axes as shown in Figure P5.5. Evaluate the integral

$$\int\limits_{(e)} \begin{bmatrix} 1 & x & y \\ x & x^2 & xy \\ y & xy & y^2 \end{bmatrix} dR$$

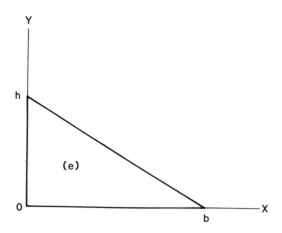

Figure P5.5 Triangular Element Positioned Along the Coordinate Axes

Result:

$$(bh/24) \begin{bmatrix} 12 & 4b & 4h \\ 4b & 2b^2 & bh \\ 4h & bh & 2h^2 \end{bmatrix}$$

5.6 Recall from Equation (5.21) that the matrix $[A]$ is defined

$$[A] = \begin{bmatrix} 1 & x_1 & y_1 \\ 1 & x_2 & y_2 \\ 1 & x_3 & y_3 \end{bmatrix}$$

Using this definition, determine $[A]$ and $[A]^{-1}$ for the triangular element shown in Figure P5.5 above. Then, by using Equation (5.57), calculate $\{g\}^{(e)}$ in terms of $\{f\}^{(e)}$ for the element of Figure P5.5. Compare the results with Equation (5.76).

Result:

$$\{g\}^{(e)} = (bh/12) \begin{bmatrix} 2 & 1 & 1 \\ 1 & 2 & 1 \\ 1 & 1 & 2 \end{bmatrix} \{f\}^{(e)}$$

5.7 Using Equations (5.73) and (5.75), verify the result in Equation (5.76). Compare the computational effort with that of Problems 5.5 and 5.6.

Section 5.10

5.8 Using the method of separation of variables solve the boundary value problem of Equations (5.92) and (5.93) with $T = C = 10$ on the Y-axis

$$\partial^2 T/\partial x^2 + \partial^2 T/\partial y^2 = 0 \quad \begin{array}{c} 0 \leqslant x \leqslant 1 \\ 0 \leqslant y \leqslant 1 \end{array}$$

where $T(0, y) = 10$, $T(1, y) = T(x, 0) = T(x, 1) = 0$. Specifically, let $T(x, y) = X(x) Y(y)$ where $X(x)$ and $Y(y)$ are functions of x and y, respectively, thus separating the independent variables. Show that this leads to the following ordinary differential equations for X and Y:

$$d^2 X/dx^2 - \lambda^2 X = 0$$

and

$$d^2 Y/dy^2 + \lambda^2 Y = 0$$

where λ is a constant. Show further that the requirement that $T(1, y) = T(x, 0) = T(x, 1) = 0$ leads to $\lambda = n\pi$ where n is a positive integer and that then $T(x, y)$ becomes

$$T(x, y) = \sum_{n=1}^{\infty} b_n (\sinh n\pi x - \tanh n\pi \coth n\pi x) \sin n\pi y$$

where the bn are constants (Fourier coefficients). Finally, show that satisfying the boundary condition $T(0, y) = 10$ leads to the solution

$$T(x, y) = \sum_{n=1}^{\infty} (20/n\pi)(1 - \cos n\pi)(\cosh n\pi x$$
$$- \sinh n\pi x \coth n\pi) \sin n\pi y$$

5.9   Using the results of Problem 5.8, show that $T(1/3, 1/3) \approx 3.818$.

5.10   Using Equations (5.23), (5.24), (5.29), (5.30), and (5.39), verify Equation (5.95).

5.11   Verify Equation (5.99).

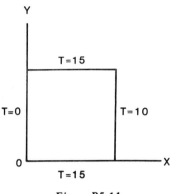

Figure P5.11

Section 5.11

5.12   Using the procedure outlined in the example problem of this section, find the interior nodal temperatures if the boundary temperatures are as shown in Figure P5.11.

5.13   From the variational principles, explain why the governing equations reduce to simply four equations: (5.106) to (5.109).

Section 5.12

5.14   Using Equations (5.120) to (5.126) verify Equations (5.127) to (5.130).

Section 5.13

5.15   Verify Equation (5.163).

5.16 Using Table 5.1 of Section 5.10, develop a typical element stiffness matrix, say $[\overset{(6)}{\Lambda}]$ for plane stress/plane strain problems.

Answer: Each incidence matrix has 6 rows and 18 columns. For $[\overset{(6)}{\Lambda}]$ all entries are zero except for the ones in columns 7, 7, 8, 8, 12, and 12 for rows 1, 2, 3, 4, 5, and 6, respectively.

REFERENCES

1. E. Volterra and J. H. Gaines, *Advanced Strength of Materials*, Prentice Hall, Englewood Cliffs, NJ, 1971.
2. Y. C. Fung, *Foundations of Solid Mechanics*, Prentice Hall, Englewood Cliffs, NJ, 1965.
3. A. C. Ugural and S. K. Fenster, *Advanced Strength and Applied Elasticity*, American Elsevier, New York, 1975.
4. R. W. Little, *Elasticity*, Prentice Hall, Englewood Cliffs, NJ, 1973.

6

Isoparametric Elements

6.1 INTRODUCTION

With this brief chapter we conclude our introduction to the finite element method with an elementary discussion of isoparametric elements. We discussed triangular elements in Chapter 5. Modeling accuracy with triangular elements can be improved by increasing the number of elements. There may be occasions, however, when it is more convenient to improve the accuracy by using different elements. For example, we may want to employ elements with curved edges to match the boundary of a given region, as shown in Figure 6.1.

One method of developing these elements is to use the so-called "isoparametric formulation." This procedure is based upon a mapping, or transformation, of a relatively simple element into the desired shape. Figure 6.2 illustrates such a mapping where a square element is mapped into a quadrilateral element. The square element is often called the "parent element" and the quadrilateral element is called the "image element" or the "parametric element." In this

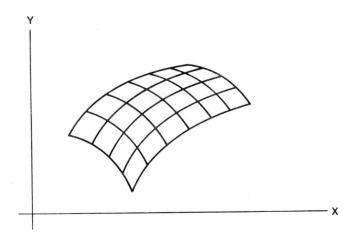

Figure 6.1   Quadrilateral Elements with Curved Boundaries

case ξ and η are the "local" coordinates and x and y are the "global" coordinates.

Parametric equations are used to relate the (ξ, η) and (x, y) coordinates. These relations may be developed by using interpolation functions as discussed in Chapter 5. If these same interpolation functions are also used to represent the dependent field variable over the image element, the element is called an "isoparametric element."

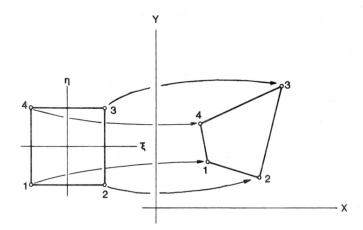

Figure 6.2   Mapping of a Square Element into
a Quadrilateral Element

In what follows we will discuss the basic concepts of this iso-parametric formulation. As before, we will restrict our discussion to simple cases remembering that the more complex cases are different primarily in complexity but not in concept.

6.2 RECTANGULAR ELEMENTS

In preparation for our discussion of isoparametric elements consider the rectangular element shown in Figure 6.3. As before, our objective is to describe a function $\psi(x, y)$ over the element with the following requirements: (1) ψ assumes the values ψ_1, ψ_2, ψ_3, and ψ_4 at the respective nodal points; and (2) ψ varies linearly along the edges of the element. Regarding notation, let

$$\psi_i = \psi(x_i, y_i) \quad i = 1, 2, 3, 4 \tag{6.1}$$

To meet these requirements we can follow a procedure similar to that of Section 5.4 and expand ψ in a polynomial in x and y. That is, let ψ be written in the form

$$\psi(x, y) = a + bx + cy + dxy \tag{6.2}$$

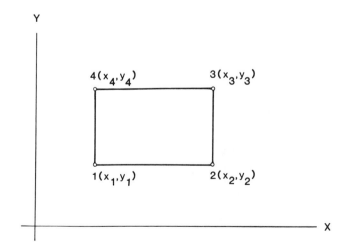

Figure 6.3 A Rectangular Element

The coefficients a, b, c, and d may be determined by requiring ψ to satisfy the requirements of Equation (6.1). That is,

$$\psi_1 = a + bx_1 + cy_1 + dx_1 y_1$$

$$\psi_2 = a + bx_2 + cy_2 + dx_2 y_2$$

$$\psi_3 = a + bx_3 + cy_3 + dx_3 y_3 \tag{6.3}$$

$$\psi_4 = a + bx_4 + cy_4 + dx_4 y_4$$

In matrix form Equation (6.3) may be written

$$\{\psi\} = \begin{bmatrix} \psi_1 \\ \psi_2 \\ \psi_3 \\ \psi_4 \end{bmatrix} = \begin{bmatrix} 1 & x_1 & y_1 & x_1 y_1 \\ 1 & x_2 & y_2 & x_2 y_2 \\ 1 & x_3 & y_3 & x_3 y_3 \\ 1 & x_4 & y_4 & x_4 y_4 \end{bmatrix} \begin{bmatrix} a \\ b \\ c \\ d \end{bmatrix} = [A]\{a\} \tag{6.4}$$

where the arrays $[A]$ and $\{a\}$ are defined by inspection. Solving for the coefficient array $\{a\}$ we obtain

$$\{a\} = [A]^{-1}\{\psi\} \tag{6.5}$$

From Equation (6.2) and (6.5), on the element, $\psi(x, y)$ may thus be written

$$\psi = [1 \ \ x \ \ y \ \ xy] \begin{bmatrix} a \\ b \\ c \\ d \end{bmatrix} = \{x\}^T\{a\} = \{x\}^T [A]^{-1}\{\psi\} \tag{6.6}$$

or

$$\psi = \{N\}^T\{\psi\} \tag{6.7}$$

where by inspection $\{x\}^T$ is defined as the row array $[1\ x\ y\ xy]$, and $\{N\}^T$ is defined as the array

$$\{N\}^T = \{x\}^T [A]^{-1} \tag{6.8}$$

Hence, on the element, $\{N\}$ may be expressed as

$$\{N\} = \{N(x, y)\} = [A^T]^{-1}\{x\} \tag{6.9}$$

By expanding Equation (6.7) we see that ψ may be written, on the element,

$$\psi(x, y) = \psi_1 N_1(x, y) + \psi_2 N_2(x, y) + \psi_3 N_3(x, y)$$

$$+ \psi_4 N_4(x, y) \tag{6.10}$$

The members of the array $\{N\}$ thus have the form of interpolation coefficients as discussed in Chapter 5. We can see this more clearly by examining Equation (6.9): By multiplying by $[A^T]$ we obtain

$$\{x\} = [A^T]\{N\} = \begin{bmatrix} 1 & 1 & 1 & 1 \\ x_1 & x_2 & x_3 & x_4 \\ y_1 & y_2 & y_3 & y_4 \\ x_1 y_1 & x_2 y_2 & x_3 y_3 & x_4 y_4 \end{bmatrix} \begin{bmatrix} N_1 \\ N_2 \\ N_3 \\ N_4 \end{bmatrix} = \begin{bmatrix} 1 \\ x \\ y \\ xy \end{bmatrix} \tag{6.11}$$

If we let x and y have the values x_1 and y_1, this becomes

$$\begin{bmatrix} 1 & 1 & 1 & 1 \\ x_1 & x_2 & x_3 & x_4 \\ y_1 & y_2 & y_3 & y_4 \\ x_1 y_1 & x_2 y_2 & x_3 y_3 & x_4 y_4 \end{bmatrix} \begin{bmatrix} N_1 \\ N_2 \\ N_3 \\ N_4 \end{bmatrix} = \begin{bmatrix} 1 \\ x_1 \\ y_1 \\ x_1 y_1 \end{bmatrix} \tag{6.12}$$

The array on the right side of this expression is thus equal to the first column of $[A^T]$. Hence, at the nodal point 1 with coordinate (x_1, y_1) the interpolation array $N(x, y)$ has the form

$$\{N(x_1, y_1)\} = \begin{bmatrix} 1 \\ 0 \\ 0 \\ 0 \end{bmatrix} \qquad (6.13)$$

Similarly we find

$$\{N(x_2, y_2)\} = \begin{bmatrix} 0 \\ 1 \\ 0 \\ 0 \end{bmatrix}, \qquad \{N(x_3, y_3)\} = \begin{bmatrix} 0 \\ 0 \\ 1 \\ 0 \end{bmatrix},$$

$$\qquad (6.14)$$

$$\{N(x_4, y_4)\} = \begin{bmatrix} 0 \\ 0 \\ 0 \\ 1 \end{bmatrix}$$

Hence, the members N_i $(i = 1, \ldots, 4)$ of the array $\{N\}$ have the property

$$N_i(x_j, y_j) = \begin{cases} 1 & i = j \\ 0 & i \neq j \end{cases} \qquad (6.15)$$

This is a property expected of interpolation coefficients. Finally, let $N_i(x, y)$ $(i = 1, \ldots, 4)$ have the value *zero* for all points exterior to the element. Thus, by examining Equation (6.9), (6.10), and (6.15) we conclude that $\{N\}$ is an array of interpolation coefficients for the rectangular element.

Finally, we see that the expression for (x, y) in Equation (6.10) meets the requirements stated earlier: That is, ψ has the values ψ_1, ψ_2, ψ_3, and ψ_4 at the nodal points 1, 2, 3, and 4, and ψ varies linearly along the edges of the element.

6.3 QUADRILATERAL ELEMENTS

Consider next the quadrilateral element shown in Figure 6.4. As before, our objective is to express the field variable $\psi(x, y)$ in terms of its values at the nodal points 1, 2, 3, and 4. Indeed, we seek a representation of ψ in the form [see Equation (6.7)]

$$\psi(x, y) = \{\psi\}^T \{N(x, y)\} \tag{6.16}$$

where $\{\psi\}$ and $\{N\}$ are the arrays of nodal values and interpolation functions respectively.

Global convergence and continuity between elements is assured for second order problems if: (1) the representation of ψ is linear along the element edges and (2) constant values of the partial derivatives of ψ can be represented over the element (see, for example, Zienkiewicz*).

In attempting to determine the interpolation functions of Equation (6.16) subject to the above requirements, we might try expressing $\psi(x, y)$ in the form

$$\psi = a + bx + cy + dxy \tag{6.17}$$

However, if we attempt to determine the coefficients a, b, c, and d from the equations

$$\psi_i = a + bx_i + cy_i + dx_iy_i \quad (i = 1, \ldots, 4) \tag{6.18}$$

where, as in Equation (6.3), x_i and y_i are the nodal coefficients, we find that the variation of ψ along the element edges is *not* linear (see Problem 6.6).

---

*O. C. Zienkiewicz, *The Finite Element Method*, Third Edition, McGraw-Hill, New York, 1977, Chapter 8.

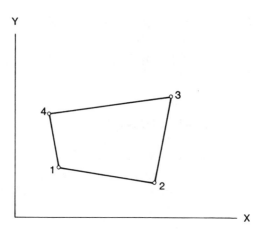

Figure 6.4 A Quadrilateral Element

To overcome this difficulty consider mapping a square "parent" element into the quadrilateral element as depicted in Figure 6.5. It is readily seen that the mapping of the square element into the quadrilateral element is given by the expressions

$$x = (\tfrac{1}{4})[x_1(1 - \xi)(1 - \eta) + x_2(1 + \xi)(1 - \eta)$$
$$+ x_3(1 + \xi)(1 + \eta) + x_4(1 - \xi)(1 + \eta)]$$ (6.19)
$$y = (\tfrac{1}{4})[y_1(1 - \xi)(1 - \eta) + y_2(1 + \xi)(1 - \eta)$$
$$+ y_3(1 + \xi)(1 + \eta) + y_4(1 - \xi)(1 + \eta)]$$

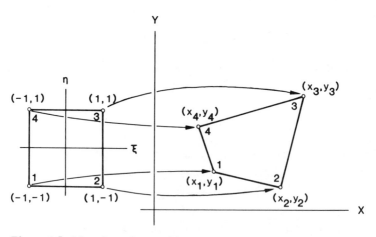

Figure 6.5 Mapping a Square Element into a Quadrilateral Element

Next, let the functions N_1, N_2, N_3, and N_4 be defined:

$$N_1 = (\tfrac{1}{4})(1 - \xi)(1 - \eta)$$

$$N_2 = (\tfrac{1}{4})(1 + \xi)(1 - \eta)$$

$$N_3 = (\tfrac{1}{4})(1 + \xi)(1 + \eta)$$ (6.20)

$$N_4 = (\tfrac{1}{4})(1 - \xi)(1 + \eta)$$

By inspection we see that these functions have the properties of interpolation functions in that

$$N_i(\xi_j, \eta_j) = \begin{cases} 0 & i \neq j \\ 1 & i = j \end{cases}$$ (6.21)

and

$$\sum_{i=1}^{4} N_i = 1$$ (6.22)

where ξ_i and η_i are the nodal coordinates in the ξ-η plane of Figure 6.5. Thus x and y of Equations (6.19) may be written in the form

$$x = \sum_{i=1}^{4} x_i N_i = x(\xi, \eta)$$

  (6.23)

$$y = \sum_{i=1}^{4} y_i N_i = y(\xi, \eta)$$

By "inverting" these expressions we could solve for ξ and η in terms of x and y. Then from Equations (6.20) we could express N_i ($i = 1, \ldots, 4$) as functions of x and y. This motivates us to express ψ in the form

$$\psi = \psi_1 N + \psi_2 N_2 + \psi_3 N_3 + \psi_4 N_4 = \sum_{i=1}^{4} \psi_i N_i = \{N\}^T \{\psi\}$$
  (6.24)

where now the functions N_i ($i = 1, \ldots, 4$) are defined to have the value zero at points exterior to the element. It is left to the reader to verify that this representation of ψ satisfies the requirements of (1) being linear along a quadrilateral edge and of (2) being able to represent constant values of $\partial\psi/\partial x$ and $\partial\psi/\partial y$ over the element.

Finally, we note that the same functions used to map the square parent element into the quadrilateral element of Equations (6.19) are used as interpolation functions in Equation (6.24). This dual use of the functions N_i is commonly called the "isoparametric representation."

6.4 ELEMENT STIFFNESS MATRICES AND FORCING FUNCTIONS

To illustrate the use of the isoparametric representation in finite element procedures consider again the two-dimensional heat conduction problem of Chapter 5. Recall that the governing system of algebraic equations for the nodal temperatures were obtained by minimizing the integral [see Equation (5.45)]:

$$I = \sum_{e=1}^{n} \overset{(e)}{I} \tag{6.25}$$

where $\overset{(e)}{I}$ is [see Equation (5.51)]

$$\overset{(e)}{I} = (\tfrac{1}{2}) \{\overset{(e)}{T}\}^T [\overset{(e)}{k}] \{\overset{(e)}{T}\} + \{\overset{(e)}{g}\}^T \{\overset{(e)}{T}\} \tag{6.26}$$

where $\{\overset{(e)}{T}\}$ is the nodal temperature array, $[\overset{(e)}{k}]$ is the element stiffness matrix, and $\{\overset{(e)}{g}\}$ is the element forcing function.

For a quadrilateral element $\{\overset{(e)}{T}\}$ is the array given by

$$\{\overset{(e)}{T}\} = \begin{bmatrix} \overset{(e)}{T_1} \\ \overset{(e)}{T_2} \\ \overset{(e)}{T_3} \\ \overset{(e)}{T_4} \end{bmatrix} \tag{6.27}$$

The procedure with quadrilateral elements thus reduces to obtaining expressions for $[\overset{(e)}{k}]$ and $\{\overset{(e)}{g}\}$ in terms of the elements' geometrical parameters. To obtain these expressions, recall that $\overset{(e)}{I}$ may be written in the form [see Equation (5.46)]

$$\overset{(e)}{I} = (\tfrac{1}{2}) \int_{(e)} [k \, (\overset{(e)}{\nabla T})^2 + 2\overset{(e)}{f} \, \overset{(e)}{T}] \, dR \tag{6.28}$$

where $\overset{(e)}{T}$ is the temperature distribution over the element. For a quadrilateral element $\overset{(e)}{T}$ may be expressed as

$$\overset{(e)}{T} = \overset{(e)}{T}(x, y) = \overset{(e)}{T_1} \overset{(e)}{N_1}(x, y) + \overset{(e)}{T_2} \overset{(e)}{N_2}(x, y) + \overset{(e)}{T_3} \overset{(e)}{N_3}(x, y)$$
$$+ \overset{(e)}{T_4} \overset{(e)}{N_4}(x, y) \tag{6.29}$$

By following the procedure of Chapter 5, let $\overset{(e)}{\nabla T}$ be written in the form [see Equation (5.37)]

$$\overset{(e)}{\nabla T} = \begin{bmatrix} \partial \overset{(e)}{T}/\partial x \\ \partial \overset{(e)}{T}/\partial y \end{bmatrix} \tag{6.30}$$

By substituting from Equation (6.29), $\overset{(e)}{\nabla T}$ may be written

$$\overset{(e)}{\nabla T} = \begin{bmatrix} \partial \overset{(e)}{N_1}/\partial x & \partial \overset{(e)}{N_2}/\partial x & \partial \overset{(e)}{N_3}/\partial x & \partial \overset{(e)}{N_4}/\partial x \\ \partial \overset{(e)}{N_1}/\partial y & \partial \overset{(e)}{N_2}/\partial y & \partial \overset{(e)}{N_3}/\partial y & \partial \overset{(e)}{N_4}/\partial y \end{bmatrix} \begin{bmatrix} \overset{(e)}{T_1} \\ \overset{(e)}{T_2} \\ \overset{(e)}{T_3} \\ \overset{(e)}{T_4} \end{bmatrix} \tag{6.31}$$

or as

$$\overset{(e)}{\nabla T} = [\overset{(e)}{B}] \, \{\overset{(e)}{T}\} \tag{6.32}$$

where by comparison of the equations $\overset{(e)}{[B]}$ is defined

$$\overset{(e)}{[B]} = \begin{bmatrix} \partial \overset{(e)}{N_1}/\partial x & \partial \overset{(e)}{N_2}/\partial x & \partial \overset{(e)}{N_3}/\partial x & \partial \overset{(e)}{N_4}/\partial x \\[2ex] \partial \overset{(e)}{N_1}/\partial y & \partial \overset{(e)}{N_2}/\partial y & \partial \overset{(e)}{N_3}/\partial y & \partial \overset{(e)}{N_4}/\partial y \end{bmatrix} \qquad (6.33)$$

By substituting from Equations (6.31) to (6.33), the first term of the integral $\overset{(e)}{I}$ of Equation (6.28) may be written

$$\int_{(e)} k\,(\nabla \overset{(e)}{T})^2 \, dR = \{\overset{(e)}{T}\}^T \left(\int_{(e)} k\,[\overset{(e)}{B}]^T \overset{(e)}{B}] \, dR \right) \{\overset{(e)}{T}\} \qquad (6.34)$$

By comparison with Equation (6.26), we see that the element stiffness matrix becomes

$$\overset{(e)}{[k]} = \int_{(e)} k\,[\overset{(e)}{B}]^T [\overset{(e)}{B}] \, dR \qquad (6.35)$$

By a similar analysis the forcing function may be written (see Problem 6.10)

$$\{\overset{(e)}{g}\} = \left(\int_{(e)} \{\overset{(e)}{N}\}^T \{\overset{(e)}{N}\} \, dR \right) \{\overset{(e)}{f}\} \qquad (6.36)$$

Equations (6.35) and (6.36) are the desired relations. A difficulty which arises, however, is the evaluation of the integrals. The reason is that the domain of integration is the x-y plane, whereas $\{N\}$ is the defined in Equation (6.19) in terms of ξ and η. Additional complications are encountered if we attempt to solve Equation (6.19) for ξ and η in terms of x and y. Therefore, we turn to numerical procedures to evaluate the integrals, as discussed in the following section.

6.5 NUMERICAL EVALUATION OF THE ELEMENT STIFFNESS MATRIX AND FORCING FUNCTION

As noted, the difficulty in the analytical evaluation of $[k]^{(e)}$ and $\{g\}^{(e)}$ stems from the integrals being taken over x and y whereas the integrands are expressed in terms of ξ and η. What is more, the matrices $[B]^{(e)}$ and $[B]^{(e)T}$ of Equation (6.35) involve derivatives of $N_i^{(e)}$ with respect to x and y.

To perform these differentiations and integrations it is convenient to introduce the Jacobian matrix $[J]$ defined

$$[J] \overset{D}{=} \begin{bmatrix} \partial x/\partial\xi & \partial y/\partial\xi \\ \partial x/\partial\eta & \partial y/\partial\eta \end{bmatrix} \tag{6.37}$$

Using Equation (6.23) $[J]$ may be written

$$[J] = \sum_{i=1}^{4} \begin{bmatrix} x_i \partial N_i^{(e)}/\partial\xi & y_i \partial N_i^{(e)}/\partial\xi \\ x_i \partial N_i^{(e)}/\partial\eta & y_i \partial N_i^{(e)}/\partial\eta \end{bmatrix} \tag{6.38}$$

From the definition of Equation (6.37), the inverse of $[J]$ becomes

$$[J]^{-1} = (1/J) \begin{bmatrix} \partial y/\partial\eta & -\partial y/\partial\xi \\ -\partial x/\partial\eta & \partial x/\partial\xi \end{bmatrix} \tag{6.39}$$

where J is the determinant of $[J]$.

To use these relations in evaluating $[k]^{(e)}$ and $\{g\}^{(e)}$ recall that the variables of integration may be changed from x and y to ξ and η through the expression*

$$dx\,dy = J\,d\xi\,d\eta \tag{6.40}$$

---

*See for example, F. B. Hildebrand, *Advanced Calculus for Applications*, Second Edition, Prentice Hall, 1976, p. 353.

Also, from the chain rule for partial differentiation* we have

$$\begin{bmatrix} \partial \overset{(e)}{N_i}/\partial \xi \\ \partial \overset{(e)}{N_i}/\partial \xi \end{bmatrix} = [J] \begin{bmatrix} \partial \overset{(e)}{N_i}/\partial x \\ \partial \overset{(e)}{N_i}/\partial y \end{bmatrix}$$

and

$$\begin{bmatrix} \partial \overset{(e)}{N_i}/\partial x \\ \partial \overset{(e)}{N_i}/\partial y \end{bmatrix} = [J]^{-1} \begin{bmatrix} \partial \overset{(e)}{N_i}/\partial \xi \\ \partial \overset{(e)}{N_i}/\partial \eta \end{bmatrix} \tag{6.41}$$

By changing the variables of integration to ξ and η by means of Equation (6.40), Equations (6.35) and (6.36) may be written

$$\overset{(e)}{[k]} = \int_{-1}^{1} \int_{-1}^{1} \overset{(e)}{[B]}{}^{T} \overset{(e)}{[B]} J \, d\xi \, d\eta \tag{6.42}$$

and

$$\overset{(e)}{\{g\}} = \left(\int_{-1}^{1} \int_{-1}^{1} \overset{(e)}{N}{}^{T} \overset{(e)}{N} \, d\xi \, d\eta \right) \overset{(e)}{\{f\}} \tag{6.43}$$

where $\overset{(e)}{[B]}{}^{T} \overset{(e)}{[B]}$ and $\overset{(e)}{\{N\}}{}^{T} \overset{(e)}{\{N\}}$ are to be expressed in terms of ξ and η. From Equations (6.33) and (6.41), $\overset{(e)}{[B]}$ and $\overset{(e)}{[B]}{}^{T}$ may be written

$$\overset{(e)}{[B]} = [J] \overset{(e)}{[\hat{B}]} \quad \text{and} \quad \overset{(e)}{[B]}{}^{T} = \overset{(e)}{[\hat{B}]}{}^{T} [J]^{T} \tag{6.44}$$

where $\overset{(e)}{\hat{B}}$ is defined

$$\overset{(e)}{[\hat{B}]} \overset{D}{=} \begin{bmatrix} \partial \overset{(e)}{N_1}/\partial \xi & \partial \overset{(e)}{N_2}/\partial \xi & \partial \overset{(e)}{N_3}/\partial \xi & \partial \overset{(e)}{N_4}/\partial \xi \\ \partial \overset{(e)}{N_1}/\partial \eta & \partial \overset{(e)}{N_2}/\partial \eta & \partial \overset{(e)}{N_3}/\partial \eta & \partial \overset{(e)}{N_4}/\partial \eta \end{bmatrix}$$

*See for example, F. B. Hildebrand, *Advanced Calculus for Applications*, Second Edition, Prentice Hall, 1976, p. 353.

Hence, Equation (6.42) may be written in the form

$$[k]^{(e)} = \int_{-1}^{1} \int_{1}^{1} [\hat{B}]^{(e)T} [J]^T [J] [\hat{B}]^{(e)} J \, d\xi \, d\eta \tag{6.46}$$

There still remains the problem of evaluating the integrals of Equations (6.42) and (6.43). Because of the algebraic complexity it is preferable to perform the integration numerically. A convenient numerical procedure, attributed to Gauss and Legendre, provides an accurate estimate of the integrals by adding weighted integral values at selected points.* Stated analytically the procedure approximates an integral I as

$$I = \int_{-1}^{1} f(x) \, dx \cong \sum_{i=1}^{n} w_i f(x_i) \tag{6.47}$$

where the w_i are weighting factors at the n points x_i and where n is the number of points selected. Accuracy of course, improves with increasing values of n. If $f(x)$ is a polynomial of degree m the representation is exact if $n \geq (m + 1)/2$. Values of x_i and w_i for small n are listed in Table 6.1.**

Table 6.1   Values of x_i and w_i

n	$\pm x_i$	w_i
2	0.57735	1.00000
3	0.00000	0.88889
	0.77459	0.55556
4	0.33998	0.65215
	0.86114	0.34785

---

*See for example, F. Scheid, *Theory and Problems of Numerical Analysis*, Schaum's Outline Series, McGraw-Hill, New York, 1968, Chapter 15.
**See for example, M. Abramowitz and I. A. Stegun, *Handbook of Mathematical Functions*, National Bureau of Standards, U.S. GPO, Washington, DC, 1975, p. 916.

For integration in the ξ-η plane, we simply integrate first with respect to ξ and then with respect to η. For elements with straight sides, sufficient accuracy is obtained with $n = 2$. Hence, in evaluating the element stiffness matrix and the element forcing function we can make approximations, as follows:

$$\int_{-1}^{1} \int_{-1}^{1} f(\xi, \eta) \, d\xi \, d\eta \cong \sum_{i=1}^{4} f(\xi_i, \eta_i) \qquad (6.48)$$

where ξ_i and η_i are

$$\xi_1 = \xi_4 = \eta_3 = \eta_4 = -0.57735$$
$$\xi_2 = \xi_3 = \eta_1 = \eta_2 = 0.57735 \qquad (6.49)$$

It can be shown, for example, that this procedure leads to exact values for the area of the element. That is,

$$\overset{(e)}{A} = \int_{e} dx \, dy = \int_{-1}^{1} \int_{-1}^{1} J(\xi, \eta) \, d\xi \, d\eta = \sum_{i=1}^{4} J(\xi_i, \eta_i) \qquad (6.50)$$

6.6 EXAMPLE: STEADY-STATE HEAT CONDUCTION IN A SQUARE REGION

We close this chapter with a simple example illustrating the use of rectangular elements. For comparison purposes we use the same heat conduction example as in Section 5.10.

Consider the square region shown in Figure 6.6. With the temperature specified on the edges, the problem is to determine the temperature distribution in the interior of the square.

As before, our procedure for obtaining a solution to this problem can be outlined as follows: First we will divide the square region into elements, thus obtaining the finite element model. Next, we will develop the element stiffness matrix. Using these results we will assemble the global stiffness matrix which in turn leads to the governing algebraic equations. Finally, we will solve these equations for the temperatures in the interior of the square region.

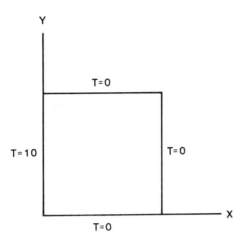

Figure 6.6 A Square Region with Specified Boundary Temperatures

To proceed, let us divide the square region into elements as shown in Figure 6.7. (This is a very coarse modeling, but it has the advantage of being simple so that the procedure may be illustrated.)

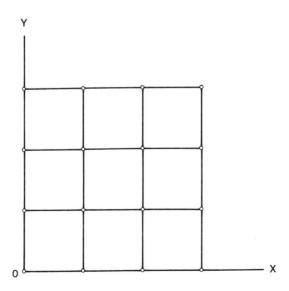

Figure 6.7 A Finite Element Model of the Square Region

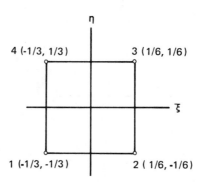

Figure 6.8   Local Coordinates of a Typical Element

Consider a typical element with its local coordinates as shown in Figure 6.8. From the results of Problem 6.3 we see that the element interpolation functions are

$$N_1^{(e)}(x, y) = (\tfrac{1}{4})(1 - 6x)(1 - 6y)$$

$$N_2^{(e)}(x, y) = (\tfrac{1}{4})(1 + 6x)(1 - 6y)$$

$$N_3^{(e)}(x, y) = (\tfrac{1}{4})(1 + 6x)(1 + 6y) \qquad (6.51)$$

$$N_4^{(e)}(x, y) = (\tfrac{1}{4})(1 - 6x)(1 + 6y)$$

From Equation (6.33) the matrix $[B]^{(e)}$ is then

$$[B]^{(e)} = \left(\frac{3}{2}\right)\begin{bmatrix} -(1 - 6y) & (1 - 6y) & (1 + 6y) & -(1 + 6y) \\ -(1 - 6x) & -(1 + 6x) & (1 + 6x) & (1 - 6x) \end{bmatrix}$$

$$(6.52)$$

Hence, from Equation (6.35) the element stiffness matrix is

$$[k]^{(e)} = k \int_{(e)} [B]^{(e)T} [B]^{(e)} \, dR \qquad (6.53)$$

or

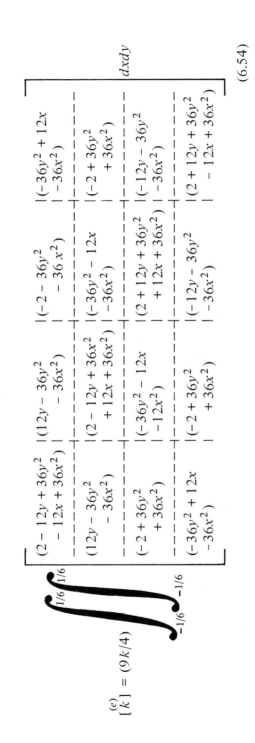

$$\overset{(e)}{[k]} = (9k/4) \int_{-1/6}^{1/6} \int_{-1/6}^{1/6} \begin{bmatrix} (2 - 12y + 36y^2 & (12y - 36y^2 & (-2 - 36y^2 & (-36y^2 + 12x \\ - 12x + 36x^2) & - 36x^2) & - 36\,x^2) & - 36x^2) \\[6pt] (12y - 36y^2 & (2 - 12y + 36y^2 & (-36y^2 - 12x & (-2 + 36y^2 \\ - 36x^2) & + 12x + 36x^2) & - 36x^2) & + 36x^2) \\[6pt] (-2 + 36y^2 & (-36y^2 - 12x & (2 + 12y + 36y^2 & (-12y - 36y^2 \\ + 36x^2) & - 12x^2) & + 12x + 36x^2) & - 36x^2) \\[6pt] (-36y^2 + 12x & (-2 + 36y^2 & (-12y - 36y^2 & (2 + 12y + 36y^2 \\ - 36x^2) & + 36x^2) & - 36x^2) & - 12x + 36x^2) \end{bmatrix} dxdy$$

(6.54)

283

By carrying out the indicated integration $\overset{(e)}{[k]}$ becomes

$$
\overset{(e)}{[k]} = k \begin{bmatrix}
2/3 & -1/6 & -1/3 & -1/6 \\
-1/6 & 2/3 & -1/6 & -1/3 \\
-1/3 & -1/6 & 2/3 & -1/6 \\
-1/6 & -1/3 & -1/6 & 2/3
\end{bmatrix}
\tag{6.55}
$$

Consider next the assembly. Let the nodes and elements be numbered and labeled as shown in Figure 6.9. For convenience in comparison and analaysis, the nodes are numbered exactly the same as in Figure 5.12 for the example of Chapter 5. In this case however there are only half as many elements.

We can now construct a table relating the element nodes to the global nodes. This leads to Table 6.2 which is analogous to Table 5.1 of Chapter 5. As before the global stiffness matrix can be constructed by inspection from the table. For example, from column (1) of Table 6.2 the entries of $\overset{(1)}{[k]}$ are inserted into the global matrix $[K]$ as follows:

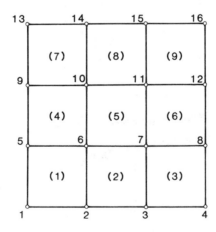

Figure 6.9 Node and Element Numbering for
the Model of Figure 6.7

Table 6.2   Matching Between Element and Global Nodes

Element Nodes	Elements								
	(1)	(2)	(3)	(4)	(5)	(6)	(7)	(8)	(9)
1	1	2	3	5	6	7	9	10	11
2	2	3	4	6	7	8	10	11	12
3	6	7	8	9	11	12	14	15	16
4	5	6	7	10	10	11	13	14	15

Global Nodes

$$k_{11}^{(1)} \to K_{11} \qquad k_{12}^{(1)} \to K_{12} \qquad k_{13}^{(1)} \to K_{16} \qquad k_{14}^{(1)} \to K_{15}$$

$$k_{21}^{(1)} \to K_{21} \qquad k_{22}^{(1)} \to K_{22} \qquad k_{23}^{(1)} \to K_{26} \qquad k_{24}^{(1)} \to K_{25}$$

$$k_{31}^{(1)} \to K_{61} \qquad k_{32}^{(1)} \to K_{62} \qquad k_{33}^{(1)} \to K_{66} \qquad k_{34}^{(1)} \to K_{65}$$

$$k_{41}^{(1)} \to K_{51} \qquad k_{42}^{(1)} \to K_{52} \qquad k_{43}^{(1)} \to K_{56} \qquad k_{44}^{(1)} \to K_{55}$$

We are now in a position to construct the governing algebraic equation. Specifically, the equation may be written in the form

$$[K] \{T\} = 0 \tag{6.57}$$

where the global stiffness matrix $[K]$ is a 16×16 array and the temperature array $\{T\}$ is a 16×1 array. From the boundary conditions of Figure 6.6, we see that the temperature is known at all but the four center nodes: 6, 7, 10, and 11. The known nodal temperatures are

$$T_1 = T_{13} = 5 \quad \text{and} \quad T_5 = T_9 = 10 \tag{6.58}$$

and

$$T_2 = T_3 = T_4 = T_8 = T_{12} = T_{14} = T_{15} = T_{16} = 0 \tag{6.59}$$

(Note the temperatures at nodes 1 and 13 are taken as the average of 0 and 10.)

The equations determining T_6, T_7, T_{10}, and T_{11} are obtained from rows 6, 7, 10, and 11 of Equation (6.57). These are

$$5K_{61} + 10K_{65} + K_{66}T_6 + K_{67}T_7 + 10K_{69}$$
$$+ K_{6,10}T_{10} + K_{6,11}T_{11} + 5K_{6,13} = 0 \qquad (6.60)$$

$$5K_{71} + 10K_{75} + K_{76}T_6 + K_{77}T_7 + 10K_{79}$$
$$+ K_{7,10}T_{10} + K_{7,11}T_{11} + 5K_{7,13} = 0 \qquad (6.61)$$

$$5K_{10,1} + 10K_{10,5} + K_{10,6}T_6 + K_{10,7}T_7 + 10K_{10,9}$$
$$+ K_{10,10}T_{10} + K_{10,11}T_{11} + 5K_{10,13} = 0 \qquad (6.62)$$

and

$$5K_{11,1} + 10K_{10,5} + K_{11,6}T_6 + K_{11,7}T_7 + 10K_{11,9}$$
$$+ K_{11,10}T_{10} + K_{11,11}T_{11} + 5K_{11,13} = 0 \qquad (6.63)$$

[These equations are identical to Equations (5.106) to (5.109) of Chapter 5.]

Using Table 6.2 and Equation (6.55) the global stiffness matrix coefficients of these governing equations are

$$K_{61} = -1/3, \; K_{65} = -1/3, \; K_{66} = 8/3, \; K_{67} = -1/3$$
$$K_{69} = -1/3, \; K_{6,10} = -1/3, \; K_{6,11} = -1/3, \; K_{6,13} = 0 \qquad (6.64)$$

$$K_{71} = 0, \; K_{75} = 0, \; K_{76} = -1/3, \; K_{77} = 8/3$$
$$K_{79} = 0, \; K_{7,10} = -1/3, \; K_{7,11} = -1/3, \; K_{7,13} = 0 \qquad (6.65)$$

$$K_{10,1} = 0, \; K_{10,5} = -1/3, \; K_{10,6} = -1/3, \; K_{10,7} = -1/3$$
$$K_{10,9} = -1/3, \; K_{10,10} = 8/3, \; K_{10,11} = -1/3, \; K_{10,13} = -1/3 \qquad (6.66)$$

and

$$K_{11,1} = 0, \; K_{11,5} = 0, \; K_{11,6} = -1/3, \; K_{11,7} = -1/3$$
$$K_{11,9} = 0, \; K_{11,10} = -1/3, \; K_{11,11} = 8/3, \; K_{11,13} = 0 \qquad (6.67)$$

Hence, the governing equation becomes, after some minor simplification,

$$8T_6 - T_7 - T_{10} - T_{11} = 25 \qquad (6.68)$$

$$-T_6 + 8T_7 - T_{10} - T_{11} = 0 \qquad (6.69)$$

$$-T_6 - T_7 + 8T_{10} - T_{11} = 25 \qquad (6.70)$$

and

$$-T_6 - T_7 - T_{10} + 8T_{11} = 0 \qquad (6.71)$$

Finally, solving these equations for T_6, T_7, T_{10}, and T_{11} leads to the results

$$T_6 = T_{10} = 35/9 = 3.88889 \qquad (6.72)$$

and

$$T_7 = T_{11} = 10/9 = 1.1111 \qquad (6.73)$$

In Chapter 5, we obtained [Equations (5.118) and (5.119)]

$$T_6 = T_{10} = 3.75 \qquad (6.74)$$

and

$$T_7 = T_{11} = 1.25 \qquad (6.75)$$

From Problem 5.8 we obtained the exact solution leading to the results

$$T_6 = T_{10} \cong 3.818 \qquad (6.76)$$

and

$$T_7 = T_{11} \cong 1.193 \qquad (6.77)$$

Thus, we see that we obtained comparable accuracy with the rectangular elements, as with the triangular elements, and that we only used half as many elements!

PROBLEMS

Section 6.2

6.1 Verify that ψ as given by Equation (6.10) varies linearly along the edges of the element.

6.2 Let a rectangular element be square and centered at the origin as shown in Figure P6.2. Find the matrices $[A]$ and $[A]^{-1}$ of Equations (6.4) and (6.5).

Result:

$$[A] = \begin{bmatrix} 1 & -1 & -1 & 1 \\ 1 & 1 & -1 & -1 \\ 1 & 1 & 1 & 1 \\ 1 & -1 & 1 & -1 \end{bmatrix} \quad [A]^{-1} = \tfrac{1}{4} \begin{bmatrix} 1 & 1 & 1 & 1 \\ -1 & 1 & 1 & -1 \\ -1 & -1 & 1 & 1 \\ 1 & -1 & 1 & -1 \end{bmatrix}$$

6.3 See Problem 6.2. Use Equation (6.8) to find the interpolation functions $N_i(x, y)$ $(i = 1, \ldots, 4)$.

Result:

$$N_1(x, y) = (\tfrac{1}{4})(1 - x - y + xy) = (\tfrac{1}{4})(1 - x)(1 - y)$$

$$N_2(x, y) = (\tfrac{1}{4})(1 + x - y - xy) = (\tfrac{1}{4})(1 + x)(1 - y)$$

$$N_3(x, y) = (\tfrac{1}{4})(1 + x + y + xy) = (\tfrac{1}{4})(1 + x)(1 + y)$$

$$N_4(x, y) = (\tfrac{1}{4})(1 - x + y - xy) = (\tfrac{1}{4})(1 - x)(1 + y)$$

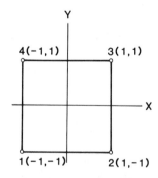

Figure P6.2 A Square Rectangular Element
Centered at the Origin

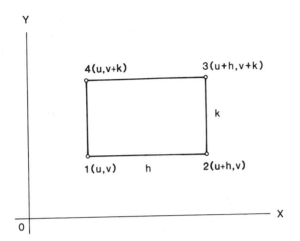

Figure P6.4 A General Rectangular Element

6.4 See Problem 6.3. Find the interpolation functions for the element shown in Figure P6.4.

Result:

$$N_1(x, y) = (1/hk)(x - u - h)(y - v - k)$$

$$N_2(x, y) = (1/hk)(x - u)(-y + v + k)$$

$$N_3(x, y) = (1/hk)(x - u)(y - v)$$

$$N_4(x, y) = (1/hk)(x - u - h)(-y + v)$$

6.5 See Problems 6.3 and 6.4. Show that $N_1 + N_2 + N_3 + N_4 = 1$.

Section 6.3

6.6 Show that Equation (6.17) does not provide a linear representation of $\psi(x, y)$ along the edges of the quadrilateral element.

6.7 Show that Equations (6.19) do indeed map the square element into the quadrilateral element of Figure 6.5.

6.8 Show that, where $\xi = \pm 1$, x and y as given by Equation (6.19) are linear functions of η and the N_i are then linear functions of x and y. Show that similar, analogous results occur when $\eta = \pm 1$.

6.9 See Problem 6.8. Show that if ψ is expressed ast $\{N\}^T \{\psi\}$ as in Equation (6.24), then ψ varies linearly along the edges of the quadrilateral element.

Section 6.4

6.10 See Equation 5.50. Verify Equation (6.36).

Section 6.5

6.11 Verify Equation (6.39).

6.12 Verify Equation (6.41).

6.13 Verify Equation (6.44).

6.14 Consider the element shown in Figure P6.14. Using Equations (6.23) and (6.37), find $J(\xi, \eta)$.

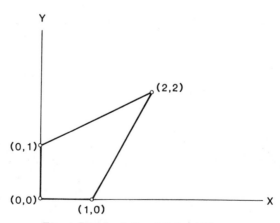

Figure P6.14 A Quadrilateral Element

Result: $J = \dfrac{1}{8} [4 + \xi + \eta]$

6.15 Using Equation (6.50) and the result of Problem 6.14 find the area of the element of Figure P6.14.

Answer: 2

Section 6.6

6.16 Verify Equation (6.51).

6.17 Verify Equation (6.52).

6.18 Verify Equation (6.54).

6.19 Verify Equation (6.55).

6.20 Verify Equations (6.64) to (6.67).

Index